Experimental Animal Behaviour

A Selection of Laboratory Exercises

Experimental Animal Behaviour

A Selection of Laboratory Exercises

Michael H. Hansell

B.A., D. Phil.

Lecturer in Zoology
University of Glasgow

and

John J. Aitken

B.Sc., M.Sc., Dip.Ed., M.I.Biol.

Senior Lecturer in Biology
Jordanhill College of Education, Glasgow

Illustrated by *Norma Aird*
Media Resource Officer,
The Robertson Centre, Paisley

Blackie

Blackie & Son Limited
Bishopbriggs
Glasgow G64 2NZ
450/452 Edgware Road
London W2 1EG

International Standard Book Number
0 216 90325 4 pre·79 E01

591·0724 HAN

Printed in Great Britain by
Robert MacLehose, Glasgow

Preface

This book has been written to provide teachers and their students with a collection of simple carefully-tested exercises in animal behaviour. Essentially it is a source book. Teachers at a variety of levels in universities, colleges and schools should be able to dip into this volume and find exercises appropriate to their particular course requirements.

As far as possible we have tried to provide all the information necessary for running the exercises successfully in the laboratory. In almost all cases the equipment required is of a simple kind. This has the advantage of cutting costs and reducing preparation time – both of which are important considerations when coping with large classes. It also has the effect of making the exercises suitable for students with limited experience of advanced techniques. The species have been selected to provide representatives from a wide range of animal classes, but our selection has been restricted to those animals from which speedy and satisfactory results can be obtained under normal teaching conditions. All the animals are readily available but, if difficulty is experienced in obtaining a particular species, the exercises described will serve as models for similar work on a related species, which may well produce interesting comparative results.

The first part of the book deals with the problems which confront a student undertaking an investigation into some aspect of an animal's behaviour, from the moment of introduction to the animal to the communicating of the findings of the investigation in the form of a written report. It offers suggestions on how these problems should be tackled, under the following main headings:

1. Observing and scoring behaviour
2. Experiments and experimental design
3. Analysis of data
4. Presentation of results
5. Writing a laboratory report

In the second part, each of the sections deals with a single animal species, their order of presentation being a conventional phylogenetic sequence. For each species, at least one, and sometimes four or five exercises are described. If practical exercises are being sought, not for a particular species of animal but to illustrate a particular feature of behaviour (e.g. orientation, learning, social behaviour), this classification is provided in the index.

The section devoted to each animal commences with a brief description of its general biology and information on how to obtain and maintain it. The practical exercises for that species are then described, and the section ends with a reference list which allows the main points raised by the practical exercises to be pursued in the original literature, either by teacher or student. Each of the exercises is described in a standard format under the following headings.

Purpose: A statement of the value of the exercise as an illustration of a feature of the behaviour of animals.

Preparation: A list of any aspects of the exercise which must be given attention before the students can commence the work, together with information on how long before the start of the exercise they must be completed (one day, one week, etc.).

Materials: A shopping list of all pieces of equipment required.

Methods: A description of how the exercise should be conducted, together with comments on problems that may arise, and a discussion of results.

Time: An estimate of the time required to carry out the exercise.

An exercise may be a simple demonstration of some aspect of animal behaviour, taking 5 minutes, or a systematic experimental investigation, taking as many weeks. In a number of instances, a simple demonstration is recommended for young students, whereas an experimental investigation of the same problem is suggested for more advanced students.

M. H. HANSELL
J. J. AITKEN

Acknowledgments

The problem of acknowledgment is a very real one. While many variations in technique may be innovations on our part, it must be made clear that we make no claim ourselves to the originality of the experimental topics. That is not the purpose of the book. Thus we wish to record that we are indebted to the many workers, some known to us but many unknown, who are the true originators of much of the experimental work collected here. We have tried as far as possible to indicate sources and authorities in the text as appropriate.

We should also like to acknowledge our debt to Eleanor Brazil, Saroj Datta, Janice Elder, Dr. Leonie Ewing, Dr. David Fraser, Dr. Donald Graham, Dr. Felicity Huntingford, Patricia Monaghan, John Simmons and Keenan Smart for valuable information and advice on specific exercises. Our thanks are due to Dr. Harry Charlton and Morag Taylor for permission to use their photographs. We would like to express our particular appreciation to Professor Aubrey Manning for some very helpful comments on certain sections of the manuscript. We are grateful to Dr. Thomas R. Bone and Professor D. R. Newth for facilities at Jordanhill College of Education and the University of Glasgow.

We are indebted to the undergraduate students at the University of Glasgow, the University of Khartoum and Jordanhill College of Education, whose laboratory courses have been the proving grounds for many of these exercises.

Contents

PART A

Introduction to Experimental Procedures

1. Observing and scoring behaviour

What is behaviour?

Any definition is problematical, but behaviour is usually taken to mean the observable movements of part or the whole of an animal's body in response to internal or external environmental factors. Most frequently it is what we can perceive of an animal's reaction to a change in environmental circumstances. On the other hand, some ethologists are interested in the occurrence of rest periods or sleep: so that even 'doing nothing' may be a useful class of behaviour for some purposes. The key to successful behaviour study, however, lies in accurate observation. The student must therefore acquire this essential skill through careful training.

Someone new to the subject, observing a fly rubbing its legs together, may state baldly that the fly is 'doing nothing'; or on hearing one chick making soft, pleasure twitters and another making piercing distress calls, may record both birds as 'cheeping'. Such lack of perception and discrimination are common, and underline the importance of an initial thorough training in observation.

Dividing up the behaviour

Once the student knows what behaviour is, and is conscious of the need for accurate observation, the problem of dividing up what can be virtually a continuous behaviour record into meaningful and useful units of behaviour can then be tackled. This is probably best approached by simple observation of the animal or animals in a situation where they have a number of behavioural options. For example, ten minutes spent watching a fly in an observation chamber with a lump of damp sugar, and perhaps other flies, should produce 'feeding', 'walking', 'flying', 'standing' and 'preening'. These are meaningful units because

> (a) they are distinct from one another;
> (b) they appear to serve different functions.

The next step would be to take one of these units, e.g. preening – and break it down further into smaller discrete behavioural bits or units. An hour's careful study of preening may produce a dozen or more such units: 'back legs together', 'back legs over wings', 'back legs under wings', etc. These are all meaningful units in the above sense, but are they useful? There are two important guides as to whether or not the behaviour units are useful:

> (a) Is one behaviour unit invariably followed by another?
> (b) What do I want to find out after I have classified the behaviour?

As an example of (a), consider the characteristic call of the male cuckoo (*Cuculus canorus*), which has two distinct components, 'kuk' and 'koo'. The one invariably follows the other. It is not useful to treat them as separate units, because any experimental manipulation which alters the occurrence of one will almost certainly similarly alter the occurrence of the other.

In the case of (b), 'the extension of the right prothoracic leg' may be a useful unit of behaviour to someone interested in the walking patterns of insects, but for someone interested in the social behaviour of insects, the movements of all legs can usefully be lumped together as 'walking'.

Arbitrary scales

It is sometimes desirable to divide one type of behaviour, using a scale to indicate degree, quantity or intensity of the behaviour. For example, an experimenter might devise a five-point scale for nest-building in mice:

> 5 points = tidy compact nest, with roof and plenty of cotton wool
> 1 point = untidy loose nest, with no roof and small amount of cotton wool.

The division of the scale is, of course, completely arbitrary and depends on the requirements of the observer. The scale could be divided into only four or into twelve (see page 17).

Naming the behaviour pattern

When the behaviour has been divided up into units, there comes the problem of what to call each unit. This is an important stage, since misnaming may lead later to misunderstanding. If a day-old chick is placed alone in a box, it will begin to emit rhythmic loud cheeps – a clearly recognizable unit of behaviour. This behaviour might be given any of the following names:

> 1. loud cheeps
> 2. mother-attracting call
> 3. distress call.

The first of these simply describes what is *observed*: the chick is observed to cheep loudly. The second indicates what *function* is ascribed to the call: that of attracting the mother hen to the chick. The third describes what is deduced to be the animal's internal state which *caused* it to call. Whereas the first name is based simply on what was observed, the functional and causal names are *interpretations* of what was observed. Names 2 and 3 are not justified until experiments have been conducted to establish, for example, if the call does attract the mother hen. The naming of the behaviour units at the end of the initial observation period should indicate simply what was observed. Each name should, at the same time, be accompanied by a complete definition of that behaviour unit in terms of what was observed.

Anthropomorphism

An experimenter's initial hypothesis may be based on all kinds of experience: experience of that animal, other animals, or even of humans. When starting the study of animal behaviour, the dangers of anthropomorphism, i.e. of interpreting an animal's behaviour in terms of human behaviour or emotions, is emphasized. This is a very necessary warning, as it is a common failing. An initial hypothesis may be founded on the similarity between the behaviour of humans and that of the animal, where no other comparison is within the experience of the experimenter: loud cheeping from an isolated chick certainly sounds like calling for the mother, in the same way as does a baby's crying. However, this does not allow the interpretation that loud cheeping *is* a call to attract the mother: it simply provides a hypothesis which can be experimentally tested.

Recording behaviour

In order to collect the results of an experiment, two basic types of equipment are needed:

> 1. a recording device
> 2. a timing device.

Together these allow a record to be made of what the animal did and when. In the simplest case, the recording device is a piece of paper and a pencil, and the timing device is a clock or watch.

From the behaviour recorded, two types of experiment can be distinguished. In the first type, the behavioural response is recorded only once – at the *end point* of the experiment, which is a fixed interval after the application of the stimulus (e.g. recording the number of maggots which have burrowed after two minutes in the dark). The second type of experiment is one in which a record is made of the behaviour *throughout* the experimental period. This may involve the recording of the occurrence of six or more types of behaviour for several minutes. Some preparation by the student is necessary, but useful information can be collected using paper, pencil and stopwatch. Before using the equipment, however, the student must decide whether he wants to score

> (a) the number of times each behaviour occurs, or
> (b) the amount of time spent doing each behaviour.

If the former is to be scored, a data sheet with a column for each type of behaviour is needed, and a tick is made in the appropriate column whenever a new behaviour starts. The total observation time might, of course, be divided up into smaller sections of, say, one minute, and the number of each behaviour class occurring in the first, second, etc., minute scored throughout the observation period.

If the experimenter wishes to score the time spent doing each behaviour, using only pencil and paper recording, the best method is by *time sampling*. This involves scoring at, say, five-second intervals the behaviour which the animal is showing. With such a small time interval, however, it is not possible to look at the watch and the animal, and to put a mark in the appropriate behaviour column. A simple alternative is to use the time interval provided by the ticking of a metronome. This works quite well when observing invertebrates, but animals such as chicks or mice will be disturbed by the ticks. Electronic audio timers can be made for a few pounds, and the experimenter can listen to these with a transistor earphone to avoid disturbing the animals. This time sampling method, especially if the time intervals are small, gives a fairly accurate measure of the proportion of time spent at each behaviour.

There are experiments involving continuous observation, where information is required not only on the number of times each behaviour occurs but also on the duration of each bout, and even the sequence of behaviour patterns. To get this information, it is really necessary to have a special recording device, such as one with a moving chart paper and a multi-channel trace to show the onset and end of each behaviour. The observer here just presses an array of six or eight buttons, each one specific to one type of behaviour. In such machines recording and timing devices are combined, and the rate of passage of the paper is constant and known. The cassette tape recorder embodies the same principles, and is very useful in a field situation where both hands may be occupied in holding binoculars or where data sheets may blow away. The smoked drum of a kymograph may be used in a similar way to record the activity cycles of an animal throughout the day. Here only two kinds of behaviour are scored — activity and inactivity — but it has the virtue of allowing results to be collected without the need for the experimenter to be present continuously for

several hours or days! It is a simple automatic recording device.

Cine film and videotape can conveniently be used to record behaviour at more advanced levels. Videotape has the virtues of not requiring high light intensity (which disturbs animals), of having instant and repeated playback, and of having reusable tape. However, the method of recording allows only a very poor and rather distorted image to be obtained, by stopping the tape in one position. Cine film, on the other hand, while not having the convenience of videotape, does permit detailed analysis of a film, frame by frame, with a special projecor.

Birds, wasps, termites and many other animals build nests; certain marine polychaetes lay down tubes; caddis larvae construct houses; pond snails leave tracks in mud; and there are even fossil tracks left by ancient animals, both vertebrate and invertebrate. All these are behavioural records. Some animals can be persuaded to produce such a record under laboratory conditions, and there are strong reasons for using such techniques with beginners, even when they are not essential to the experiment in progress. The more visible and tangible a behavioural record is, the more valuable it becomes to teacher and beginner alike.

Animal tracks can be recorded in a number of ways, e.g. smoked glazed paper — more familiar, perhaps, in the context of kymograph work — is a most sensitive and valuable device for recording the tracks of small non-aquatic animals. Records of the 'righting' behaviour of animals as small as woodlice are left quite clearly on smoked paper, as are the tracks of small light invertebrates such as millipedes and centipedes. Another method, reasonably well known, is the 'development' of mucous trails of animals such as snails or planarians, by sprinkling talcum powder on the glass on which the animals have been moving. The talc adheres to the mucous but not to clean glass. Thus when the glass is swilled gently with water, the snail or planarian track stands out clearly as a white line. The recording and interpretation of tracks left by animals in sand or snow is covered fully in a delightful book by Ennion and Tinbergen (1967), called *Tracks*.

Duration of experimental period
It would be helpful to give suggestions on how long an

experiment or set of observations should be continued, but this depends very much on the animal and on the problem. For example, the response of mosquito larvae to movement can be read three seconds after presentation of the stimulus, while cycles of activity in chicks require observation periods of a couple of hours. The duration of an experiment must be estimated during the initial phase of observation so that it can be long enough to be useful, without being excessive.

2. Experiments and experimental design

Apart from contributing to a better understanding of animal behaviour, practical work in the subject teaches students the principles of scientific inquiry through the design and execution of experiments, and the analysis and interpretation of their results. This section points out the essential features of experimental design which should be brought to the attention of students when they first start to conduct their own experiments. Section 3 considers the analysis of the results obtained.

Familiarity with the animal
Before any experiment is carried out on an animal, the student experimenter should have some familiarity with the animal to be used. This familiarity should take the form of

(*a*) direct personal experience of the animal through observation and handling;
(*b*) background information obtained from the teacher or through reading.

At the most introductory levels, time may be so limited that personal experience of the animal may be no more than the 15-20 minutes of a class, whereas for higher teaching levels, and in project work, many hours of observation and classification of the animal's behaviour may precede the first experiment. Similarly, the amount of background information may be variable, ranging from a few minutes' introduction by the teacher (or half a page of worksheet), to the reading of relevant books and original articles.

Formulating the problem
The first requirement for the design of the experiment

itself is for the student to decide what *hypothesis* the experiment is designed to test. This is very important because there is a great temptation at all levels to conduct experiments to *see what happens if...*. The experiment should be devised in such a way that the hypothesis is either confirmed or rejected on the basis of its results. Students should write down briefly but precisely the hypothesis and the title of the experiment, before anything else. In this way they are obliged to be clear about the reason for the subsequent design of the experiment.

The hypothesis may be very simple. For example, blowfly larvae (*Calliphora*) placed on peat are observed to burrow down. It is decided to test the influence of light on the burrowing rate of blowfly larvae or maggots. The title of the experiment will therefore be

Experiment to test the influence of light on the burrowing rate of blowfly maggots.

The hypothesis to be tested could be that light will make maggots burrow either faster or slower than darkness; however, it is usual to formulate a hypothesis of *no difference* between experimental and control groups. This is a *null hypothesis,* which it is usually hoped that the experiment will allow to be rejected. In the example here, the null hypothesis is that maggots will burrow *equally fast* in light and dark conditions.

Reading the literature
At the level of advanced project studies, an examination of the literature in the area planned by the experiment should be undertaken before the experiment is

carried out. At undergraduate levels this is difficult or impossible without some teacher guidance, but such a literature search is an important exercise in itself. The result of such a search may be to indicate the kind of experimental design appropriate to the problem being studied, or at least to influence it.

Experimental variables

In experimental procedure the word *variable* is used in a specific sense, but it should be pointed out to students that it means no more than the name implies – that it is some factor which varies. Adapting slightly the definition given by Meyers and Gossen (1974),

> A variable is some general property or characteristic of events, objects, animals, etc, that may take different values at different times.

Variables can be further divided into two classes – *independent* and *dependent* variables.

Independent variable. An independent variable is one which is systematically varied by the experimenter during the experiment.

If the hypothesis to be tested is that light does not influence the burrowing rate of blowfly maggots, then the independent variable to be systematically varied must be the light intensity.

Dependent variable. A dependent variable is that measure of behaviour on which the animal's performance is to be assessed. In the case of the maggots, it is the rate of burrowing.

Controls

It is essential to convey to students the fundamental principle of controlling those variables which might otherwise have a spurious effect on the independent variable being investigated. Of course, students cannot be expected to control variables which are unknown to them and which are outside the scope of the course. This may make the conclusions of their experiment incorrect, but they have nonetheless been taught that controls are essential, and that their conclusions, like the conclusions of any experiment, would be invalid if it were shown that some hitherto unsuspected and uncontrolled variable could have influenced the animal's behaviour.

Variations among the animals to be tested

No two animals of any species can be considered identical. It is therefore necessary to ensure that the experimental and control groups do contain animals of the same degree of diversity.

An example of a failure to control for such differences between animals would be to choose from the home dish a sample of the 20 most active maggots to be placed in group A, and the next 20 most active for group B. In an experiment to test the effect of light intensity on burrowing rate, any difference in burrowing rate between the two groups might then be due not to the different light intensity but simply to the fact that group A was more active than group B.

There are three common methods used to control for the variability of animals selected for the different experimental groups:

(a) Random selection
(b) Matching
(c) Using an animal as its own control.

Random. This is the most common method and essentially means the selection of the next animal that comes to hand. In case animals which come to hand most easily should be different in behaviour from those which are picked later, individual animals should be assigned to each of the experimental groups in turn. This method is usually adequate if samples are large.

Matching. Here an attempt is made to match animals between experimental groups for particular behaviour or for other features which might act as confounding variables. For example, in a study of acquisition of bar-pressing by rats in a Skinner box under two different reward schedules, each individual in the first group might be matched with one in the second group according to age – the point being that there is evidence to show that age does affect learning ability. There is no point in encouraging students to match for features which are unlikely to confound the result.

Using an animal as its own control. Here possible differences between animals in different experimental groups are controlled by using the same animals in experimental and control situations. It is, however, necessary to be sure that exposure to the first situation does not produce changes in the animal which remain to affect the measure of the dependent variable in the second situation. Alternatively, this possibility can be

Table 2.1 A common form of experimental design with experimental and control groups treated in an identical manner except for the difference of the independent variable.

Group	Selection of subjects	Independent variable	Dependent variable
Experimental	Random	A_1	D
Control	Random	A_2	D

controlled by using a *Latin square* design (Table 2.5).

Environmental variables
In a natural or semi-natural environment, features such as temperature, humidity, light intensity and time are continually changing. When the independent variable has been chosen, steps should be taken to control for other environmental variables, as far as is possible. Time cannot, of course, be controlled – it passes inexorably; however, changes in an animal's performance over time may be cyclical, e.g. circadian, so that this may be controlled by testing animals in the same part of each day.

Natural and artificial experimental environments
It is necessary for students to appreciate that any animal is a complex and sensitive object adapted for living in a particular kind of environment, and that consequently there is the danger that imposing rigid limitations in an experimental situation may produce abnormal behaviour. An alternative reason for not imposing such constraints is that certain kinds of behaviour may occur only in relatively rich and varied environments. Thus it is often difficult for ethologists to show differences resulting from different ex-

perimental treatments, because of the "noise" of uncontrolled variables. Often this problem is overcome only by testing large samples.

Statistical treatment
Where statistical treatments are to be used on the experimental data, the type of treatment planned should be considered beforehand to avoid data wastage. Analyses by students of the results of their own experiments, in whatever field of science, is an effective method of familiarizing them with certain statistical techniques. Thus if it is intended that they should become familiar with the use of the *t-test,* their samples should be large enough to allow such a treatment, and there should be only two treatment groups to allow comparison of one with the other.

Design
The most common type of experimental design is one with single experimental and control groups and a random selection of subjects, as shown in Table 2.1.

It may be desirable to test several values of the independent variable. For example, the burrowing rate of blowfly maggots may be tested at three different levels of illumination and the control group tested in complete darkness, as in Table 2.2.

Table 2.2 An experimental design in which the same response (dependent variable) is measured for 4 different values of light intensity (independent variable).

Group	Selection of subjects	Independent variable	Dependent variable
Experimental 1	Random	Bright light	Rate of burrowing
Experimental 2	Random	Moderate light	Rate of burrowing
Experimental 3	Random	Dim light	Rate of burrowing
Control	Random	Darkness	Rate of burrowing

It is not uncommon to have an experimental design which examines in a single experiment the effects of two independent variables. This is a *factorial design,* and is characterized by the presence of experimental groups for all combinations of the two independent variables. So to investigate three values of variable A under two different values of variable B, there would be six groups, as shown in Table 2.3.

Variable B

		b1	b2
	a 1		
Variable A	a 2		
	a 3		

Table 2.3 Factorial design for the examination of the effect on the dependent variable of the interaction between 3 values of independent variable A and 2 values of independent variable B, giving 6 groups altogether.

Using blowfly maggots as an example again, these empty their crops about 3 days before pupation – an occurrence which is indicated by the disappearance of a brown patch at the anterior end of the body. Maggots with empty crops migrate away from food into the ground to pupate. So an experiment could be designed to test the burrowing rates, under 3 different light intensities, of maggots with full and with empty crops (Table 2.4).

Crop state

	Crop full Dark	Crop empty Dark
Light intensity	Crop full Low light	Crop empty Low light
	Crop full High light	Crop empty High light

Table 2.4 Application of the factorial design shown in Table 2.3 to investigate the burrowing rates of maggots, at 3 light intensities with two crop states giving 6 test groups as shown.

In experiments where individual animals or groups of animals are being used as their own controls, it is possible that exposure to one experimental situation will influence performance in a subsequent situation. The effect of this carry-over may be controlled by running the sequence of experiments on successive groups in as many different orders as possible; equal numbers of animals are then exposed to each sequence of test situations. This is called a *Latin square* design.

For example, if the burrowing of maggots is to be observed in coarse, medium and fine sawdust, it might be that maggots when being tested in their third situation are more fatigued or desiccated than in the first. To control for this effect, a Latin square design is used (Table 2. 5).

	1st Test	2nd Test	3rd Test
Group 1	Fine	Medium	Coarse
Group 2	Coarse	Fine	Medium
Group 3	Medium	Coarse	Fine

Table 2.5 Latin square design to control for the effect of position in time of each of the 3 substrates to be tested. To control for the effect of specific substrates occurring before others, all possible orders of the 3 situations (i.e. 6 groups) should be run.

3. Analysis of data

Measurement

Different kinds of measurement permit only certain types of statistical treatment. These can be considered as different *levels* of measurement:

> Nominal (classificatory)
> Ordinal (ranking)
> Interval and ratio.

The ways in which data may be analysed depend upon the level of measurement used.

Nominal. This is the weakest level of measurement. Animals are classified, according to a particular characteristic, into one of a number of mutually exclusive sub-classes. For example, the circadian activity of a particular species of animal might be classified as either *nocturnal* or *diurnal*. Animals might be classified by habitat as *pelagic, planktonic* or *benthic*.

Ordinal. The classification of animals into subgroups according to some feature may be such that particular sub-classes can be said to be of *higher* or *lower* measure than other sub-classes. For example, animals might be classified into 4 groups according to activity:

> 1. Motionless
> 2. Slightly active
> 3. Active
> 4. Very active.

This is a four-point scale of activity giving an *order* or *rank* of sub-classes 4   3   2   1. It is not permissible to say, using such a scale of measurement that, for example, a score of 4 is twice as high as a score of 2, but only that it is a higher score. This level of measurement is particularly useful for behavioural data.

Interval and ratio. When the scale of measurement is such that values not only can be ranked according to size but also bear an exact relationship to their distances along the scale, then the scale is an *interval* scale. If, in addition, the scale has a true zero point, it is a *ratio* scale. In a ratio scale, the ratio between two measures on the scale is independent of the units used to measure it. So, for example, the ratios between 30 and 60 mph and between 44 and 88 ft/s are both 2. Both scales have interval measurement and a true zero.

Parametric and non-parametric tests

The first introduction to statistical procedures for determining levels of significance of observed differences between experimental groups tends to be through tests such as the *t*-test. This is a *parametric test* and, together with other parametric tests, has certain disadvantages which do not arise in non-parametric tests. Non-parametric tests are more attractive at introductory levels because

> (*a*) they may be used on small samples;
> (*b*) some of them require only very simple computation.

They are more attractive for behaviour data because

> (i) they do not assume normally distributed data;
> (ii) they may be used on data no better than those of ordinal or nominal measurement, whereas parametric statistics may be applied only to data of interval or ratio measurement.

These features together make non-parametric statistical tests particularly suitable for introducing to students the concepts of null hypothesis, probability and significance.

Detailed information on the appropriate use of non-parametric tests, together with a clear explanation of each of a number of non-parametric tests, may be found in Siegel (1956).

4. Presentation of results

Tables

Tables are a convenient way of presenting information where several sets of data concerning several animals or situations have to be shown. Tables are effective provided that they are not composed of too many columns and rows. A table with ten columns and ten rows is, after all, giving the reader 100 items to associate and compare with one another. It is, unfortunately, not uncommon to find in published literature tables of twice that size. There may be good reasons for this – such as to need to demonstrate dominance relationships by presenting all the threats given and received by whom to whom within a group of 20 animals. However, if a table is going to be a large one, the student should consider more economic or graphic ways of presenting the information – such as by histogram or graph or, indeed, a simplified table. Table 4. 1 shows the features of a well-presented table: it is hardly necessary to say that the paper is on the food-finding orientation of an aquatic snail, since that is clear from the table alone.

The table is good because

(a) its headings are brief and clear;
(b) it is composed of only a few columns and rows;
(c) there is a single line at the bottom which summarizes the main finding illustrated by the table.

The data collected by a student during a particular laboratory or project may not be amenable to such succinct presentation, but this table shows the main points to be considered.

Graphs

If there is a choice between summarizing the results in the form of a table or of a graph, almost invariably a

Table 4.1 The proportion of snails reaching the source within 10 minutes (from Townsend, 1973)

| | Lettuce extract source | | Control source | |
	Number tested	Number reaching source	Number tested	Number reaching source
Series A	8	7	8	1
Series B	10	8	10	2
Series C	15	12	15	4
Series D	20	14	20	7
Overall	53	41	53	14

graph is better, because it makes apparent the essential findings at a glance. However, before resorting to a graph, it is important that students should understand the conventions. A graph is used to present the change in value of one parameter resulting from the change in value of the other. In a laboratory experiment, the value of one of these parameters is altered by the experimenter. It is therefore *independent* of the animal being observed (an *independent variable*). The value of the other parameter is dependent on the nature of the response of the animal (a *dependent variable*). The independent variable is conventionally represented along the horizontal or *x*-axis (axis of abscissae), and the dependent variable along the vertical or *y*-axis (axis of ordinates).

Frequently the *x*-axis measures 'time', but it could be temperature, generations, etc. Figure 4.1 illustrates the weight (mass) of a nest built by a mouse over a period of several days.

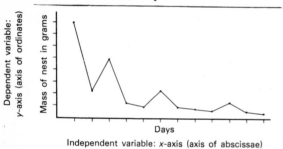

Figure 4.1 The change in mouse nest weight over days showing the independent variable on the axis of abscissae (*x*-axis) and the dependent variable on the axis of ordinates (*y*-axis).

It is important to note that, as in this figure, the points on a graph should be joined up by a series of straight lines. There is a strong tendency among students to draw 'best-fit' straight lines or elegant curves 'by eye': this is incorrect (figure 4. 2). There are formulae to determine 'best-fit' straight lines or the

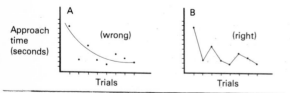

Figure 4.2 The time taken for one chick to approach another in a series of approach trials. A – Shown with a curve drawn incorrectly through the points by eye; B – Shown drawn correctly with a straight line connecting the neighbouring points.

shapes of curves; these may be calculated at more advanced levels if desired.

Commonly the same experimental procedure is applied to more than one group of animals, in which case direct comparisons may be made by plotting all the results on one graph. However, even when different types of lines and points are used to represent the different groups, for reasons of clarity it is not advisable to plot more than three lines on one graph (figure 4. 3) – unless, perhaps, they show no overlap, when more may be included without affecting the clarity of the presentation (figure 4. 4).

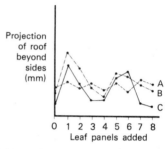

Figure 4.3 The projection of the roof of a caddis house in front of the sides against number of roof panels replaced in 3 different test groups A, B and C.

Although the lines are marked in a different pattern, the trends are not immediately apparent because of the overlap of the 3 lines.

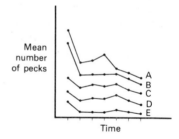

Figure 4.4 Number of pecks given over time for 5 test groups of chicks, A – E. The clear separation of the lines allows trends to be seen clearly on a single graph.

The *y*-values represented on a graph are usually *mean* values for the group: they do not, therefore, give any impression of the variation or range of the sample upon which that mean value is based. The variation, however, can be conveyed on the graph by the inclusion of bars to indicate the standard deviation or standard error of the *y*-values about each point (figure 4. 5). The calculation of standard deviation is unlikely to be an expected part of writing a report, except at the more advanced levels. While at less advanced levels, a bar indicating the *range* of values may be used, this can be misleading, as it places the emphasis on the extreme upper and lower values. A much better, yet very straightforward, method of presenting the variation of the sample is to plot each point as a *median* value (the value on either side of which lie equal numbers of observations), then divide each half of the range above and below the median into halves of equal length to find the *quartile points*. The bars obtained by joining the quartile points represent the *interquartile range* (Bishop, 1968).

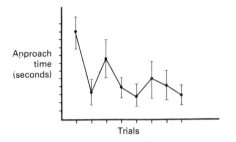

Figure 4.5 Time taken for one chick to approach another in a series of approach trials. Points represent mean approach time per trial. Vertical bars through each point represent ± one standard deviation either side of the mean.

Histograms

Histograms also provide a way of clearly presenting data which might otherwise have been displayed less clearly as a table. A histogram is used when the quantity or intensity of a given behaviour is represented on the *y*-axis, while the *x*-axis contains different classes. These classes may be arranged in any order, depending on what the histogram is intended to convey (figure 4. 6).

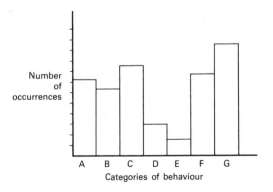

Figure 4.6 Histogram showing the mean number of occurrences of 7 different categories of behaviour, A to G.

y-values are usually mean values, as represented by the heights of the columns, and the variation about a mean may be indicated by the inclusion of a standard-deviation or standard-error bar in the same manner as for graphs. Again, as in the case of graphs, data from two or more groups may be compared on the same histogram (figure 4. 7).

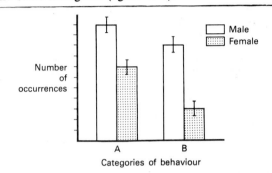

Figure 4.7 Histogram showing the mean number of occurrences of 2 different categories of behaviour (A and B) in males and females. Vertical bars indicate the standard error of each sample.

Photographs

The camera is a useful piece of equipment for recording behaviour. In a project type of study, one of the objectives may be to develop the skill of the student in using the camera as a behaviour-recording instrument. In a short laboratory class, such an objective is not feasible, but the use of a Polaroid camera, which is easy to operate and gives instant results, can provide worthwhile behavioural records. However, inexpensive Polaroid cameras, such as might be used in large classes, are rather limited in performance. They will not give a detailed picture of anything much smaller than a rat and, even then, only if the animal is practically stationary.

Photographs may be used as sources of behaviour information which must then be assessed and analysed. An example of this would be the measurement of individual distance within a group of animals, as a study of their social structure. Photographs taken for this purpose alone are, therefore, not appropriate to the 'Results' section, but rather to the 'Appendix'. However, it may be desirable for a student to include photographs in the Results section as the clearest way of conveying particular information – for example, to illustrate a characteristic posture adopted by an animal, which would be difficult to convey verbally

Figure 4.8 Two characteristic postures adopted by king penguins; lying on the belly and standing on the heels with toes in the air; here photography illustrates clearly characteristic behaviour difficult to convey in words.

(figure 4. 8). It is important to resist the temptation to include photographs merely for decoration, or to illustrate points that would be more clearly shown in some other way.

Line drawings

A photograph is often not as good at illustrating a particular behaviour as a simple line drawing from which all unnecessary detail has been excluded. Line drawings may be drawn or traced from photographs or from frames of cine film (figure 4. 9). For some purposes, however, line drawings may be drawn from life, allowing some essential impression of the animal's behaviour to be conveyed in the report, where the use of the camera is impracticable or uneconomic.

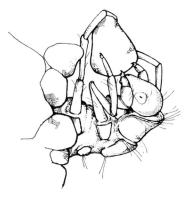

Figure 4.9 Line drawing of a caddis larva manipulating a sand grain taken from a not very good video-tape. The use of the line drawing gives greatly improved clarity.

Other types of presentation

Pie charts. Where a group of observations is composed of a number of sub-groups, then the group as a whole can be represented by a circle, and the sub-groups by segments of the appropriate size (figure 4. 10).

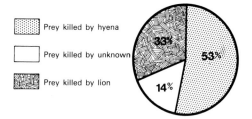

Prey killed by hyena

Prey killed by unknown

Prey killed by lion

Figure 4.10 A 'pie' chart showing the percentage of prey dying in three different ways when lions and hyenas were both seen feeding off the same carcass (after Krunk, 1972)

Kite diagram. A kite diagram is essentially a histogram with the bars indicating the size of the

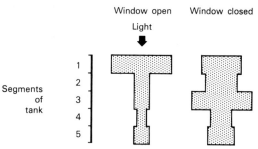

Figure 4.11 Kite diagrams to show the distribution of the crustacean *Corophium volutator* in an experimental tank with lateral light (left) and without lateral light (right) (after Barnes *et al.*, 1969)

measurement aligned across a central axis. This kind of diagram is often used to represent the distribution of animals in space, and is particularly effective in comparing results from more than one situation (figure 4. 11).

Orientation diagram. The orientation of animals at a point, or their direction of movement from that point, can be recorded as a series of wedge-shaped bars radiating from the point, where the length of each bar is proportional to the number of animals orientating in that direction (figure 4. 12).

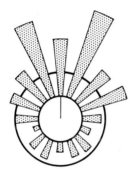

Figure 4.12 Diagram to show the orientation of manx shearwaters on release in sunny conditions. The direction of home is indicated vertically and the length of each bar is proportional to the number of birds leaving in that direction (after Matthews, 1955).

Maps. In field studies or studies of social behaviour, it is often desirable to include maps to show, for example, the distribution of animals' burrows or territorial boundaries (figure 4. 13).

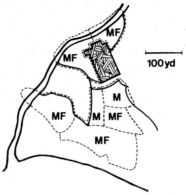

Figure 4.13 A map showing the distribution and size of robin territories in the proximity of a house. M = territory of a male; F = territory of a female; MF = territory of a pair (after Lack, 1943).

Traces. Where smoked-paper traces are a convenient size for incorporation into a report, they should be evenly varnished, trimmed to frame the part of the trace required, and mounted on card or stiff paper. Frequently it is important to put a fully labelled or annotated line diagram alongside the varnished paper.

Smoked papers which are too bulky for inclusion, even in an appendix, can be photographed. This also applies to animal tracks on glass or on other substrates which cannot conveniently be preserved and presented in any other way.

Other methods. There may be instances where an original type of diagrammatic presentation is devised by a student. This is quite legitimate, and the only criterion for judging it should be its effectiveness in comparison with other methods of presentation.

5. Writing a laboratory report

The writing of a formal laboratory report as such at lower levels in schools is rarely necessary or even desirable. Frequently animal behaviour is studied briefly in ecological contexts, or as part of a more general introduction to experimental methods. Simple observational and elementary experimental approaches to the locomotion and feeding of such animals as the common earthworm feature early in some syllabuses. Here the animals are not being investigated in a narrow behavioural sense. The objectives may be very different and much wider in scope. At this level it may be more important that the young pupil handles and works with the whole living organism than that he learns formal experimental techniques.

It is generally agreed that with young pupils it is important to

    (a) develop their powers of observation;
    (b) develop their techniques of recording their observations through succinct and unambiguous language;
    (c) initiate some sort of simple quantitative approach – counting, timing, measuring – so that techniques of handling data can be developed along simple lines.

It is not necessary to involve the pupil in writing up an experiment if the above three objectives can be more economically achieved through the use of a well-designed worksheet. Here the pupil's attention can be guided to the behaviour for investigation: well-designed questions can at once guide and test his powers of observation, and invite responses which are descriptively accurate and unambiguous. Further questions may be incorporated to ensure that, by counting and measuring, the pupil obtains data which are then available for treatment.

One of the chief drawbacks with any worksheet system is, of course, the fact that the pupil may go away with an inaccurately completed and largely valueless piece of paper. It is vital that a second sheet, or other device, be available for the child to take away which contains the essential results of the laboratory work plus further explanatory and background information.

Many teachers prefer a more flexible approach, which will involve the pupils in some sort of 'write-up'; but the principles are the same.

At the top of the school, formal records of experimental work are more valuable; and if the pupils are to develop their own patterns of experimental work along open-ended lines, it is essential that a standard format for the presentation of results be adopted. Ideally, the format chosen should mirror the methods usual at higher academic levels.

The format for the writing of a scientific paper is now fairly standard. It has developed into this format so that the order and the method of presentation of the material allow the easiest possible assimilation of the information. The reasons for this method of presentation apply equally to the writing of laboratory reports, but with the additional reasons that most of the students will be starting to read original scientific papers and some will eventually write papers of their own.

The section headings for a laboratory report should be as follows:

Introduction. In this section the nature of the problem to be investigated should be explained. The problem may be raised by previous observations or by experiments performed by the investigator or by someone else. The relevant findings of these previous studies should be mentioned briefly, and references to already published work should be accompanied by the name of the author and date of publication: the complete reference should then be given in the 'References' section.

Methods. This section should contain all the information about experimental procedure. It may very well include a line drawing of the experimental apparatus or test situation, which may be diagrammatic or representational – but a diagrammatic figure is often better, even allowing that many people cannot draw. However, a drawing should be included only where it is necessary; if it is included, there should be no unnecessary detail. This section should also contain information on such procedural details as experimental temperatures, lighting, length of test, observation and scoring systems.

Results. This section contains the results of the experiment or, more often, a summarized version of the results or data collected during the experiment. The 'raw' data are usually too bulky to be included, but may be added in an Appendix. The summarized data may be presented in a number of ways (graphs, histograms, etc.), already covered in Section 4 (p. 19). The method of presentation chosen should be

(*a*) appropriate to the data;
(*b*) the clearest possible method of presenting the data.

At more advanced levels and where time is available, it is important to perform statistical tests on the data to establish the significance of any differences observed between experimental and control groups. In this case it is necessary to state which significance test was used, and the probability level associated with the results (see Section 4, p. 19). If a specialized test is used, then the source reference should be included in the text. The section should end with a brief statement of the conclusions drawn from the observed result.

Discussion. This section is used for the interpretation of the conclusions of the previous section. It is therefore the only section which contains any speculation on the results. The section should not, however, contain any new data. It should be used to relate the findings of the experiment to other relevant work. This means that work cited in the Introduction may well be included again, since hopefully the question originally posed has at least been partially answered. New questions may have been raised relating to further work, which should be mentioned and the connection discussed.

Summary. This section should itemize in brief sentences the main findings and conclusions contained in the Results and Discussion sections.

References. This should simply be a list of names of any authors mentioned in the text of the report, together with the date of publication, title of article and name of journal for each reference, such that it would allow anyone reading a particular reference to locate the original paper.

Appendix. This section should contain all the 'raw' data collected in the experiment, the analysis of which appears in the Results part of the report. The separation of the raw and analysed data allows the reader to see the uncluttered findings in the Results and then, if he wishes, to go to the original figures in the Appendix.

Reports with more than one experiment

In situations where two or more experiments are similar in overall pattern but with minor individual differences, a general Introduction and Methods should be given. The headings then should be

Experiment 1 Details of method
Results
Experiment 2 Details of method
Results

The Discussion section should, however, be common to all the experiments, thereby drawing together the findings of the whole report.

Descriptive reports

Some behaviour investigations are not experimental in approach but descriptive. The report should, however, be written under the same main headings. The Results section is not divided up into experiments but into different themes which suit the findings of the experimenter. For a social behaviour study, the subheadings of the Results section might read:

Results: (*a*) dominance order
 (*b*) mating behaviour
 (*c*) territories
 (*d*) care of young

REFERENCES

Bishop, O. N. (1968), *Statistics for Biology,* Longman, London.
Ennion, E. A. R. and Tinbergen, N. (1967), *Tracks,* Oxford.
Meyers, L. S. and Grossen, N. E. (1974), *Behavioural Research; Theory,* W. H. Freeman, San Francisco.
Siegel, S. (1956), *Non-parametric Statistics for Behavioural Sciences,* McGraw-Hill, New York.

PART B

Laboratory Exercises

1. Turbellaria

INTRODUCTION

The Turbellaria are essentially free-living platyhelminthes. They are found in fresh water and marine environments, and some have invaded the land. Terricolous triclads, often brightly coloured, occur particularly in damp forest areas. Many genera of use in the laboratory, such as *Dendrocoelum,* have a worldwide distribution.

They are particularly suitable for introductory studies. They are attractive and robust laboratory animals, easily procured and maintained; they are harmless, slow-moving and relatively easily handled by the beginner.

The collection of fresh-water turbellarians is simple. A clean white plastic bucket should be almost filled with a mass of pond or stream plants and debris. The weed and debris should be just covered with water and left to stand. The flatworms emerge from the weed and debris and collect in the top few centimetres of water in the bucket. They can be seen easily against the white plastic and can be removed to Petri dishes with a wide-mouthed pipette.

In many instances it will be possible and preferable to collect, sort and use fresh materials. If the flatworms have to be maintained over a long period, it is probably best to sort them into species and to keep them in separate white-enamel surgical trays. In general it is safer to use rain water or water from their original habitat, rather than water from the laboratory tap. The trays should be darkened with suitable covers and kept in a cool place. The animals' food requirements will vary a little from species to species. Our cultures are largely maintained on chopped *Tubifex* or *Enchytraeus* worms, but species such as *Dendrocoelum lacteum* require live food such as *Daphnia* or juvenile amphipods and isopods.

The eight species of triclad Turbellaria likely to be found and used in the laboratory in Britain can be identified from Mellanby (1963).

EXPERIMENTAL WORK

1.1. Locomotion in turbellarians

Purpose
This simple introductory exercise is designed to draw the students' attention to the modes of locomotion in these flatworms and to give them practice in observing, handling and recording techniques.

Preparation
None

Materials
Per student:
1. Petri dish with 6 planarians, e.g. *Polycelis nigra,* or mixture of species as desired.
2. Tray of pond water, about 30cm × 30cm
3. Sheet of clean glass, 25cm × 25cm
4. Container of powdered talc
5. 2 cavity slides
6. 4 cover slips
7. Low-power stereomicroscope (× 20 or × 40)
8. Tube of carmine powder
9. Access to high-power monocular microscope
10. Wide-mouthed pipette
11. Fine soft paint brush
12. White tile

Methods
The initial observations on the mode of locomotion of the flatworms provided can be made as the animals move about the Petri dish. The animals are first either contracted, rounded-up and quiescent, or partially extended but quiescent. They tend to rest more extended

in the angle of the dish, being more likely to contract and round up in the open.

When the flatworms move, three types of movement can be distinguished:

(*a*) Head-waving
(*b*) Alternately extending and contracting (earthworm-like)
(*c*) Gliding

(*a*) Head-waving is most frequently seen when the animal is up against a barrier, or when it is in very shallow water, or out of water. Students should pipette a specimen on to a tile and observe it under the stereomicroscope. The smaller the drop of water surrounding the flatworm, the more frequent the exploratory head-waving behaviour.

(*b*) Flatworms show their characteristic and apparently effortless 'gliding' when submerged. They extend fully and glide forward with their heads just clear of the substrate. This should be observed under a low-power stereomicroscope, by careful handling of the Petri dish, so that the flatworm is held in the field of view. The smooth gliding is accomplished by cilia on the undersurface of the animal in particular, and the currents set up by them can be detected by carmine or even by some of the fine silt which may be present in the pond water. The name of the group, in fact, is taken from the Latin *turbella,* meaning a small disturbance.

(*c*) Gliding takes the animal smoothly and evenly across a variety of substrates under water, including the underside of the surface film. The gliding is aided by the secretion of a fine carpet of mucus, along which the flatworm travels.

The presence and the action of the cilia require a study of the animal in a drop of water with carmine particles, on a cavity slide and under a monocular microscope of suitable power.

The mucus carpet can be demonstrated quite easily by allowing the flatworm to glide on a sheet of glass submerged in a tray of pond water. The glass should be removed from the water and sprinkled with very fine powdered talc. The excess talc can be gently swilled away, leaving the mucus tracks showing up clearly with the white talc adhering to them. This talc 'developing' technique is a simple and useful one for any animal that leaves a mucus trail.

Time
30 minutes

1.2. The light responses of *Polycelis nigra*

Purpose
Ullyot (1936) analysed the responses of *Dendrocoelum lacteum* to light. Fraenkel and Gunn (1961) have examined Ullyot's work and developed at some length the concept of klinokinesis response, or change in the frequency or amount of turning per unit time, dependent on intensity of stimulation.

Klinokinesis, it was claimed, could explain the distribution of *Dendrocoelum* in the gradient of light intensity produced by an overhead source of illumination alone. This work seems to have focused attention on klinokinesis – a fact which has been reflected in the choice of practical exercises on the light responses of flatworms; however Stasko and Sullivan (1971) in a critical reappraisal of Ullyot's work concluded that the distribution of *Dendrocoelum* species in an experimental light gradient could adequately be explained as a directional response to small amounts of laterally scattered light rather than klinokinesis. The conditions for a klinokinetic response to a light gradient can seldom exist in nature anyway; horizontal light components must nearly always be present, and perhaps the most ecologically relevant response to light for the student to start with is the taxis.

This exercise is designed to demonstrate the response of *Polycelis* to directional light with the simplest of apparatus.

Materials

Per student:
1. Petri dish containing 10 *Polycelis nigra* (or alternative species, e.g. *D. lacteum*)
2. White tile
3. Wide-mouthed pipette
4. Bench lamp

Methods
A dozen or so drops of water should be streaked across the tile, forming a long narrow very-shallow pool. The ten flatworms should be introduced carefully to this shallow streak of water. Incident light from a window or bench lamp should be approximately at right angles to the line of water.

In most cases, there is an obvious movement directly away from the source of illumination as soon as the flatworms begin to move. The animals will push out of the restricted pool of water and travel across the dry tile, pulling fingers of water with them. After a minute or two, reverse the tile, or turn it through 90°, and again there should be an immediate change of direction by all the flatworms so that again they travel away from the light.

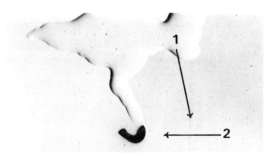

Figure 1.2.1 Turbellarian changing direction of movement in response to a change in direction of the incident light (1. initial light direction; note position of water streak left by the moving animal. 2. New light direction.)

The restricted water supply seems to accentuate the response to light. Although they will move away from light in deeper water in a tray or dish, the pattern is not so clear-cut as on the tile. The student will also notice 'gliding' in the main streak of water, then the 'inching' type of locomotion as it pushes out onto the dry tile and, finally, 'head waving', which is prominent when the tile is turned so that the incident light is coming from another direction.

If required, the light directions can be recorded on the tile and the residual mucus track 'developed' with talc.

Time
20 minutes

1.3. Rheotaxis in *Crenobia alpina*

Purpose
The tendency of stream-dwelling animals, both vertebrate and invertebrate, to head upstream is well known. Koehler (1932) investigated the responses of *Planaria (Crenobia) alpina* to moving water with the use of squirts of water from fine pipettes. This type of approach is readily reproducible, makes a fine demonstration of a positive rheotaxis, and permits a certain amount of variation and creativity in the

pattern of investigation. This exercise is designed to introduce the student to the behaviour and to the rather neat and simple technique with which to investigate it.

Preparation
Ideally the *Crenobia* should be tested beforehand to make sure that they are showing the response required. These flatworms are often kept in cold aerated tanks and may not show the positive rheotaxis until the temperature is raised above 12°C.

Materials

Per student:
1. Selection of glass tubing
2. Selection of pipette bulbs (large)
3. Dish of cool river water
4. Lump of Plasticine
5. Bunsen or equivalent burner
6. Container with 4 or 5 *Crenobia*
7. Asbestos square

Methods
Basically the method involves creating a current of water and noting the reaction of the flatworm to it. The students should make up a number of pipettes — one wide-mouthed for lifting the flatworms, and a number of very fine pipettes for producing fine streams of water. The student should first test the *Crenobia*'s response to water direction when the water in the dish is swirled gently around. The flatworms normally turn into the stream, although frequently the student finds it tricky to establish just the right speed of water. Some of the flatworms are washed off the sides of the dish and settle in the middle for a while. Eventually, it should be clear that the animals are tending to turn into the stream of slowly swirling water, particularly at the edges of the dish.

Having established that the *Crenobia* respond to moving water, the student should now find out which part of the flatworm's body is sensitive to the stream. Here the carefully controlled stream from a finely drawn pipette can help to shed light on the problem.

The water in the dish must be still at the start of the tests. The student should experiment (and possibly discuss and plan his test pattern with a fellow student or tutor) with fine streams of water directed at different parts of the body of the animal. When eventually a suitably gentle and even stream can be produced and accurately directed at the animal's head, middle and tail, the experimenter will find that stimulation of the head will produce an immediate turning towards the source of the water. It looks as if the receptors are located only at the anterior end. In fact, they are not. If the anterior end is cut off and tests are carried out before regeneration begins, the rheotaxis occurs as before (Fraenkel and Gunn, 1961). The student will also note that when stimulation is stopped or suddenly reduced, 'head waving' is a characteristic response.

Time
30-60 minutes

1.4. The location of food by turbellarians

Purpose
The use of fresh meat, liver and crushed invertebrates as bait for flatworms in rivers and ponds is well known. The aim of the present exercise is to introduce the student to chemolocation in pond or stream turbellaria, and to expose him to an area of study suitable for open-ended investigation.

Preparation
Collect and starve for a few days a number of pond planarians such as *Dugesia lugubris*.

Materials

> Per student or pair of students:
> 1. Container of 10 flatworms
> 2. Tray of pond water
> 3. 3 Petri dishes
> 4. Wide-mouthed pipette
> 5. Container with various baits (liver, etc.)
> 6. 3 small specimen tubes
> 7. Cotton wool
> 8. Capillary tubing

Methods
Ideally the laboratory exercise should follow a field introduction to the problem of food location by the flatworms. The students should have had the opportunity to put out baits and glass jars/traps with baits in them. In streams, the field worker will get results fairly quickly; in ponds, the flatworms normally take longer to find the baits. Still-water baits should be left for several hours if necessary. Jars baited with beef or liver can be left out overnight.

From the success of the trapping, it is obvious that the flatworms can somehow locate the food. The problem can then be formalized for analysis in the laboratory. The simplest test to start with is the obvious one of putting a small piece of freshly-cut liver or crushed water snail into a shallow tray of water, and closely observing the path taken to it by a starved planarian. The student should quickly identify the 'head waving' and general excited turning of the animals, and finally the direct and accurate 'gliding' straight to the bait. Koehler (1932) describes the behaviour of *Planaria (Dugesia) lugubris* when locating a source of blood from slices of the pond snail *Planorbis*. The flatworm will follow the end of a capillary tube full of a suspension of *Planorbis* blood or macerated ox liver if the tube is put close to the head of the flatworm and drawn slowly just ahead of it. This, and the nature of the path taken towards the bait in the open tray, suggest a tactic response to a chemical gradient of sufficient steepness. The student should then think in terms of quantifying the approach, timing the flatworm on the different stages of its route to the bait, varying the strength of stimulus, comparing the effectiveness of different baits, and comparing different species, e.g. *P. nigra* with *D. lugubris*.

N.B. Food is an attractant. What about a dilute solution of sulphuric acid? Here the student will find that the flatworm turns directly away from the stimulus.

Time
about 1 hour

REFERENCES

Fraenkel, G. S. and Gunn, D. L. (1961), *The Orientation of Animals,* New York: Dover.
Koehler, O. (1932), Beiträge zur 'Sinnesphysiologie der Susswasserplanarien', *Z. vergl. Physiol.,* 16: 606-756.
Mellanby, H. (1963), *Animal Life in Fresh Water: A guide to Fresh-water Invertebrates,* 6 ed., London, Methuen.
Stasko, A. B. and Sullivan, C. M. (1971), 'Responses of Planarians to Light: An Examination of Klinokinesis', *Anim. Behav. Monogr.,* 4 (2): 124 pp.
Ullyot, P. (1936 a), 'The Behaviour of *Dendrocoelum lacteum.* I. Responses at Light-and-dark Boundaries', *J. exp. Biol.,* 13: 253-264.
Ullyot, P. (1936 b), 'The Behaviour of *Dendrocoelum lacteum* II. Responses in a Non-directional Gradient', *J. exp. Biol.,* 13: 265-278.

2. The Earthworms

INTRODUCTION

Of the two commonest genera of earthworms in the British Isles, *Lumbricus* and *Allolobophora,* the former is perhaps more commonly studied. *L. terrestris* inhabits soils which do not dry out, which are not too acid, and which contain sufficient organic detritus for food. They form burrows through the soil, partly by pushing earth aside and partly by ingesting it. The burrows are usually lined with a smooth cement of defaecated earth. Burrows can extend downwards for several feet, and often terminate in a small chamber. The deepest burrows are formed in cold or dry weather, where there is a risk of frost or desiccation in the topmost layers of soil.

An earthworm normally comes to the surface in moist and dark conditions to feed. The worm's tail is anchored in the entrance to its burrow while searching on the surface for leaves and vegetable debris, so that it can retreat quickly into its burrow if disturbed. Leaves may be pulled into the burrows for safer feeding, and it was Darwin who first observed that leaves were dragged in apex first.

Reproduction begins about March when the temperature starts to rise; however, in some *Allolobophora* species, reproduction is suspended during the relatively dry summer months from July to September. Mating in *Lumbricus terrestris* takes place at the soil surface at dawn, but other species mate below ground (Edwards and Lofty, 1972).

The collection of worms by digging is reasonably simple (if sometimes arduous in dry weather or during winter), but biological-supply houses and fishing-tackle shops do keep stocks; and it is easy to set up and maintain a culture if necessary.

The worms should be kept in large wooden boxes, fibreglass tanks or similar containers filled with moist (not wet) loamy soil and partially decayed leaves to a depth of at least 30 cm. To prevent evaporation, each container should be covered with a sheet of glass or polythene. The worms must not be overcrowded. About 20 worms per $30cm^3$ soil is maximum. Pick out any damaged or limp specimens, and replace with healthy worms.

For most laboratory work, such long-term culturing is not necessary. Earthworms can be stored for several weeks in damp *Sphagnum* moss. A large all-glass tank (e.g. an aquarium tank) should be filled with layers of moss, and the worms spread on the surface. The animals slowly work their way down through the moss to the bottom of the tank. Every day or two the moss must be taken out of the container, the worms collected, the moss repacked in the tank and, finally, the worms spread on the moss surface again. The worms will again work their way down through the moss and, in doing so, become cleaner and tougher and generally more pleasant to handle during experimental work.

Worms from both suppliers and personal cultures should be put through such a cleaning, conditioning and toughening regime in the moss tank for two or three days before any experimental work.

EXPERIMENTAL WORK

2.1. Control of locomotion

Purpose

The locomotion of the earthworm takes the form of a series of waves of extension and contraction passing backwards along the animal. These movements are simple and repeated, and therefore easily observed. They illustrate a type of locomotion available to animals without legs or backbones, and the control of this locomotion may be investigated experimentally in a manner similar to the investigation of insect locomotion described in Section 10.2 (page 84). To enable students to design sensible experiments and to understand the conclusions to be drawn from those experiments, it will be necessary for them either to have dissected the nervous system of an earthworm or at least to have had a lecture on its anatomy.

Preparation
None.

Materials

Per pair of students:
1. Two good-sized earthworms (10-15 cm)
2. A stereomicroscope
3. One pair of fine scissors
4. Two pairs of watchmaker's forceps (about size 5)
5. A block of unpolished wood, 20 cm × 5 cm surface and about 1 cm thick
6. One enamel dish, approximately 20 cm × 25 cm
7. Three or four elastic bands about 2 mm or 3 mm wide
8. 15 cm of cotton thread

Methods
Observation. One worm should be placed in the enamel dish and its locomotion behaviour observed. The anterior portion of the worm will be seen to make short side-to-side and forward 'exploratory' movements, which may be followed by a wave of extension and then contraction, starting at the front end and passing smoothly along the whole length of the body, thereby drawing the animal forwards (figure 2.1.1). One wave of extension, followed by contraction, usually passes along the body before a new one is initiated. Similar waves may at times pass along the animal from posterior to anterior, drawing the worm backwards, which may well be seen when the worm is

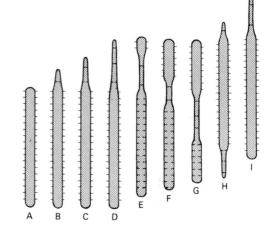

Figure 2.1.1 Diagram to illustrate the waves of extension passing back along the body of an earthworm and resulting in forward progression, A – I. Short hair-like extensions represent the setae in extended or retracted position (after Chapman and Baker, 1964)

first placed in the dish. In case of confusion, the anterior is the more pigmented end, and the clitellum, when apparent, is in the anterior half (figure 2.1.2).

Figure 2.1.2 An earthworm securely held down to a wooden block with a broad elastic band. Note the position of the clitellum towards the anterior end of the animal.

If the worm is tapped lightly on the anterior end, there will be a sudden shortening of the body. This 'escape withdrawal' response is due to the firing of the median giant fibre, which runs the length of the nerve cord. A light tap on the posterior end produces a similar sudden body contraction, caused by the firing of the pair of lateral giant fibres which lie either side of the median one but conduct from posterior to anterior. (Laverack, 1963, summarizes the evidence for this.)

Experiment. Having observed normal locomotion and escape withdrawal, students are in a position to consider how waves of contraction are initiated and propagated.

(a) Is the presence of the cerebral ganglion necessary for the initiation of waves of contraction?
(b) Is a wave of contraction propagated entirely through the CNS, or does sensory feedback from contracting segments influence the propagation of the wave?

As is appropriate to the class, these questions may be posed, and students allowed to design their own experiments to test hypotheses, or experiments may be suggested to them. In either case, the students will have to consider what may be deduced from the results of the experiments.

To determine if feedback from muscle contraction is necessary for wave propagation and escape withdrawal. Ideally one would wish to cut through the whole body wall leaving the two ends of the worm joined only by the nerve cord. This is difficult to achieve without breaking the worm completely into two parts; however, an approximation to this condition may be achieved by fastening the worm to a block of wood by an elastic band placed across its middle (figure 2.1.2). The band should be sufficiently tight so that the worm may not slip under it but not so tight as to cause injury. An unpolished wooden block is used in preference to a plastic box lid, for example, simply because it has an unslippery surface, thereby preventing the worm from squeezing under the elastic band. The surface area of the block should be such that the worm may attempt normal locomotion on its surface without falling over the edge, since this disturbs normal contraction waves.

It will be seen in this experiment that waves of contraction initiated at the anterior end normally pass backwards until they reach the elastic band, and then appear uninterrupted at the other side and pass on to the posterior tip. Assuming that muscular contraction of the 2 or 3 segments under the band has been prevented (this may not be strictly true), the experiment demonstrates that a normal wave of locomotion may be propagated without peripheral feedback. It does not demonstrate, as students may at first believe, that peripheral feedback has no influence on the propagation of waves of locomotory movement; indeed, the experiments described below demonstrate this influence.

While the worm is secured to the wooden block, the

conduction of the escape withdrawal across the restraining rubber band may be studied by lightly tapping first the anterior end and then, after a pause of about 2 minutes, the posterior end. The withdrawal response in this situation will be found to be rather poor, possibly because the restraint of the rubber band induces vigorous attempts at locomotion which inhibit the escape withdrawal; however, if while one end is tapped the opposite tip of the animal is observed, a small contraction will be seen to occur instantly. The scale of contraction beyond the elastic band will be of the same order as that before it, indicating that neither the passage nor the intensity of the signal was affected by the elastic band.

To determine if conduction through the ventral nerve cord is necessary for wave propagation and escape withdrawal. For this investigation it is necessary to cut through the ventral nerve cord or, preferably, to remove a section of the cord from 4 or 5 segments. As there is no very satisfactory way of anaesthetizing worms for this operation without impairing their subsequent locomotion, it is unfortunately necessary to restrain unanaesthetized animals for the operation. This is done by fastening the worm *ventral side up* on the wooden block by two elastic bands, placed only about 6 cm apart, with the section of worm from which the cord is to be removed lying between the two bands. It is preferable to remove the cord about two thirds of the way along the body, for a reason which is described below.

The restrained worm should be placed under a dissecting microscope, which allows the position of the nerve cord to be seen through the body wall, as indicated by the dark line of the sub-neural blood vessel attached to its ventral surface. The cord is then exposed by a scissor-cut through the body wall alongside the cord for about 5 segments. When the cord is fully exposed by the scissors, a section of it should be removed by means of the fine forceps. This necessitates the removal of that part of the sub-neural blood vessel, but the loss of blood is slight.

When the worm is released, it will be seen that when it moves over a slippery surface, such as the wet floor of its enamel dish, waves of contraction passing along the body stop at the cut, the part posterior to the cut simply being dragged along passively. If, however, the worm is placed on a less slippery surface, such as a piece of paper, the wave propagates across the cut apparently as normal. This suggests that segments behind the cut which are denied a command through the CNS are unable to initiate a contraction wave unless the segments immediately posterior to the cut are first sufficiently stretched. However, if the cut is fairly posterior and the floor of the dish is slippery, the friction of the segments behind the cut is insufficient to cause a critical amount of stretch, while with a non-slip floor such as paper, sufficient stretch can be achieved.

A supplementary experiment supporting the above conclusion may be performed after conduction of the escape withdrawal response across the cut (described later) has been tried. In this experiment, the worm is cut into two separate portions at the position of the nerve cord operation. A piece of cotton thread is then tied around the anterior part of the posterior portion of the worm, which is laid on a piece of paper.

When the cotton thread is slack, the portion of worm will be motionless, but a light pull on the thread stretching the most anterior segments causes a wave of contraction to pass along the body to the posterior tip. This and the preceding experiment demonstrate that peripheral feedback is capable of inducing a wave of contraction, and therefore that such waves may be initiated at least artificially in the absence of specific commands from the cerebral ganglion.

Before cutting the worm into two separate parts, the escape withdrawal should be observed in response to light taps delivered to the anterior and posterior ends. A good response will be observed in these free-moving worms; however, it will be seen that a tap delivered to the posterior tip of the worm causes a sudden body contraction only in the part posterior to the cut, the portion anterior to the cut continuing to locomote normally. Similarly the initiation of an escape contraction in the anterior part of the worm has no effect on the part posterior to the cut. This demonstrates that the nerve cord (in fact the giant fibres) alone is responsible for propagating the sudden contraction; the conduction rate presumably is so fast in the giant fibres as to render feedback modification of the signal impossible.

The second worm is provided so that operations which may have been technically deficient or results which were indefinite may be repeated and improved upon.

Time

2-2½ hours is needed to complete the work described in this section. Additional time is necessary to discuss the conclusions of the experiments, particularly if students are designing their own.

2.2. Light-sensitive areas and the response of the worm to light

Purpose

Earthworms spend most of their lives in the earth, emerging only rarely at dawn or dusk to mate or to pull leaves into their burrows. This avoidance of strong light intensities suggests that the worms should be able to detect and respond to light stimuli, even though light-sensitive organs are not externally visible. The present exercise investigates the response to light and, in conjunction with the paper of Hess (1925), which gives the histological evidence of the light-detecting abilities of the earthworm, it can be used to discuss the value of behavioural measures as assays of sensory capabilities.

Preparation

Bench lamps or microscope lamps will have to be modified by the attachment of a fibre-optic light-guide.

Materials

Per pair of students:
1. Two earthworms, 10-15 cm long
2. One enamel dish, approximately 20 × 25 cm
3. A 60-watt bench lamp or microscope lamp with 50 cm of a 1-mm diameter single-strand optic-fibre light-guide attached to it.
4. A screening box about 35 × 25 × 20 cm deep (adapted from a cardboard box).

Methods

The enamel dish is placed within the box screen, so that it is shielded from direct sunlight. Room lights should be subdued or off (figure 2.2.1). A worm should be placed in the dish and allowed to move freely. When the worm is moving, the light guide may be used to illuminate only the prostomium of the worm. The worm will respond by recoiling from the illuminated patch. This demonstration is, in fact, not so clear if the worm is moving forwards in a 'determined'

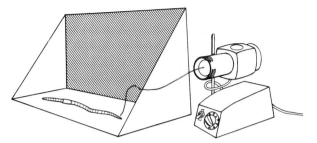

Figure 2.2.1 Earthworm screened from direct light tested for its response to light shone on a segment of the body by means of a light guide attached to a lamp.

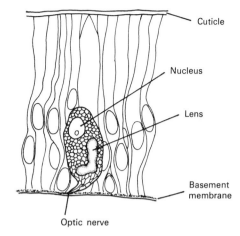

Figure 2.2.2 A single-celled light receptor of the type found in the epidermis of the earthworm *Lumbricus terrestris* (after Hess, 1925)

manner, nor when it is settled motionless in a shady or damp corner; it is most convincing when the worm is preparing to move and is making little exploratory probes with its anterior segments. When a worm is in this 'susceptible' state, it is sometimes possible by directing the light beam constantly on to the prostomium to initiate a whole series of waves of contraction, driving the animal backwards. Similar but less dramatic withdrawal may be obtained by shining the light beam on to the anterior six or so segments, but avoiding the prostomium. There does, however, appear to be not only an effect of the worm's 'mood' on. its readiness to respond, but also a change in response due to habituation to the light stimulus; persistent shining of the light in one particular area results eventually in a loss of all response.

Areas all along the body should be tested systematically, with an interval of about 2 minutes between each scan. In the middle regions of the body, it is often difficult to distinguish any response to the light from normal locomotion. It will be seen that the only response of any reasonable consistency is that achieved by shining a light on to the anterior six or so segments – particularly the prostomium. No obvious effects are achieved by illuminating other parts of the body.

Hess (1925) found single-celled light receptors (as judged by the presence of a lens) embedded in the epidermis of the worm (figure 2.2.2). These receptors

Table 2.2.1 Numbers of epidermal light receptors in the dorsal, ventral and lateral areas of the segments indicated (after Hess, 1925). The areas sampled are shown in figure 2.2.3. Body segments, except for those at the extreme anterior and posterior, contain practically no receptors.

	Median dorsal	Median ventral	Lateral	Anterior tip	Total
Prostomium	18	5	12	22	57
Segment 1	10	4	12		26
Segment 2	5	0	5		10
Segment 3	3	0	4		7
Segment 4	0	0	1		1
Ante-penultimate segment	1	0	1		2
Penultimate segment	1	0	3		4
Last segment	4	2	8		14

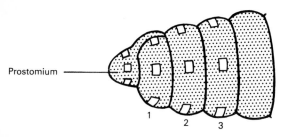

Prostomium

1   2   3

Figure 2.2.3 The anterior 3 complete body segments plus the prostomium of the earthworm *Lumbricus terrestris* seen from the side. Squares indicate the dorsal, ventral and lateral areas sampled to give the numbers of epidermal light receptors indicated in Table 2.2.1 (after Hess, 1925)

The behavioural results therefore broadly confirm the histological evidence, except that the former proves to be a less-sensitive measure — not being able to demonstrate reliably the increase in light sensitivity in the terminal segments, for example.

Time

35-40 minutes. Additional time is required for discussion to compare the behavioural with the histological evidence.

were found in sample areas of all segments — particularly on the prostomium and on the anterior complete segments, but in a rapidly decreasing number. In the middle body segments, very few were present, but the number increased somewhat again in the posterior segments (figure 2.2.3 and Table 2.2.1).

REFERENCES

Edwards, C. A. and Lofty, J. R. (1972), *Biology of Earthworms*, London, Chapman and Hall.
Hess, W. N. (1925), 'Photoreceptors of *Lumbricus terrestris* with special reference to their Distribution, Structure and Function', *J. Morph*, 41: 63-95.
Laverack, M. S. (1963), *The Physiology of Earthworms*, Oxford, Pergamon Press.

3. Tubifex

INTRODUCTION

The genus *Tubifex* is one belonging to a large family of aquatic oligochaetes, the Tubificidae. The species *Tubifex tubifex* is about 30 mm long and pink in colour due to a respiratory pigment. It has numerous segmentally arranged chaetae in dorsal and ventral bundles. It is a characteristic species of muddy and polluted fresh water, where it may occur in huge numbers. The worms burrow into the superficial layers of the mud, lining their burrows with mucus to prevent their collapse. When undisturbed, a worm has the anterior part of its body buried in the mud, while the posterior part projects upwards into the water where it undulates rhythmically to maintain a flow of the poorly oxygenated water over the body surface (figure 3.0.1).

Polluted streams and ponds are sufficiently abundant to make collection of *Tubifex* very easy. The worms can be found in the mud in shallow water, sometimes in conditions where no life seems possible. The mud can simply be scooped up in a bucket with some water. When the mud settles in the laboratory the moving tails of the *Tubifex* may be seen protruding from the mud. The worms will survive under these conditions for many days without any attention at all. To obtain *Tubifex* worms without their substrate it is easier to buy them direct from an aquarist shop, rather than go through the tedious business of separating them from the mud. A solid mass of worms like this should be kept in a shallow dish with tap water flowing slowly in one end and out the other in order to keep the water sufficiently oxygen rich.

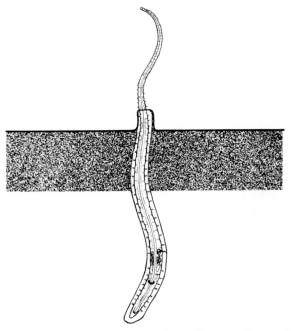

Figure 3.0.1 *Tubifex.* Side views of a tubifex worm in normal position in the mud. The anterior end has buried in the mud in a mucus-lined tube which may project slightly above the substrate; the posterior end projects into the water where it undulates back and forth in respiratory movement.

EXPERIMENTAL WORK

3.1. Study of the rate of respiratory undulation or body wriggling rate of *Tubifex*

Purpose
To determine whether the wriggle rate of individuals remains constant over time, and if the wriggle rate is related to the amount of animal protruding from the substrate.

Preparation
The mud containing the worms should be put into the specimen tubes at least 3 hours before the start of the class, to allow the worms to settle down and the water to clear.

Materials

> Per pair of students:
> 1. Four specimen tubes (2.5 cm × 7.5 cm) each containing mud with *Tubifex*
> 2. A stop watch.

Methods
Choose a slow wriggling worm and time how long it takes to produce 50 cycles of undulation (wriggles); this is easier than measuring the number of cycles per minute, which necessitates keeping one eye on the watch and the other on the worm. Make a note of the position of the worm in the tube before timing a fast worm, and a worm of intermediate speed (over 50 cycles). Having measured the rates of 3 to 5 worms, return to the first worm and time it again. Time each of the worms about 5 times over a period of 10 minutes or more. It will be found that, although the rates of different individuals is markedly different (from 25 to 65 seconds for 50 wriggles), the wriggle rate of an individual worm is very constant over time. This suggests that the oxygen requirements of individuals remains constant, at least over short periods of time.

In order to determine if there is a difference in the wriggle rate between those worms with a long projecting tail and those with a short projecting tail, all worms can be divided into 3 groups: *long, medium* and *short* according to the length of projecting tail. The wriggle rates (in the form of number of seconds to complete 50 cycles) should be measured for 5 or more animals in each of the three classes. It will be found that the 'long' group wriggles slowest and the 'short' group fastest. It is hard to be sure how to interpret this

result, and this should emerge in discussion. It could be that long tails belong to the longest worms, which happen to have a slower wriggle rate than smaller worms. Alternatively, worms with short projecting tails may be feeding at a greater depth in the mud, and need to wriggle faster to obtain the same gas exchange as a shallower-feeding worm can achieve with a longer tail at a slower wriggle rate.

Time
35-60 minutes.

3.2. Changes in wriggle rate and tail length due to increase in temperature or lowering of oxygen tension

Purpose
It may be thought that lowering of oxygen tension or increase in CO_2 causes worms to project a greater length of tail into the water and increase the wriggle rate.

We might also predict that an increase in water temperature by increasing metabolic rate and consequently oxygen requirements would also produce a similar effect. This exercise allows the above predictions to be tested.

Preparation
The mud containing the worms should be put into the specimen tubes at least 5 hours before the class to allow the worms to settle down and the water to clear. In the corked tubes it also allows the oxygen tension to drop.

Materials

> Per pair of students:
> 1. Four specimen tubes (2.5 cm × 7.5 cm) each containing mud with *Tubifex*.
> 2. Three specimen tubes (2.5 cm × 7.5 cm) as above but firmly corked and turned upside down.
> Per class:
> 3. A source of warm water (about 35°C).

Methods
To observe the effect of lowering the oxygen level (or raising the carbon dioxide level) on the respiratory behaviour of the worms, a comparison should be made between worms in corked and uncorked tubes in respect of the length of projecting tail and the wriggle rate.

First the number of worms in the *long, medium* and *short*-tailed groups in the two situations should be counted. It should be found that there is a greater proportion of worms in the *long*-tail category in the corked tubes, where oxygen levels are presumably reduced.

To compare the wriggle rate between the two situations, the time taken to perform 50 cycles should be measured for about 10 individuals from each situation, making sure to measure individuals of all tail-length groups for both situations. It should be found that there is an increase in wriggle rate shown by worms in the corked tubes.

In order to determine the effect of heating the water, two of the four uncorked tubes should be placed in a beaker of warm water (about 35°C) for 5 minutes. A comparison should then be made between worms in the warmed and unwarmed tubes. It will be found that worms in the former will project more from the mud and wriggle at a faster rate than the latter. It can be brought out in discussion that these changes in warm water may not solely be due to an increase in the animals' metabolic rate; the efficiency of the respiratory pigment has probably been depressed, and the partial pressure of the oxygen in the water reduced.

Time

From 15 minutes as a simple demonstration to 90 minutes with systematic measurement and discussion.

3.3. Responses to stimuli and habituation

Purpose

Fish such as the 3-spined stickleback (*Gasterosteus*) eat *Tubifex* worms; it is therefore reasonable to assume that a worm buried in the mud is safer from fish predation than one with a length of tail exposed. This experiment tests whether the worms have methods of detecting and responding to the approach of a predator.

Many animals show a temporary loss of a response with repeated stimulus presentation, thereby preventing the animal from responding to unimportant stimuli. This *habituation* of a response can be quickly and easily demonstrated or investigated in this exercise.

Preparation

The mud containing the worms should be put into the specimen tubes at least 3 hours before the class to allow the worms to settle down and the water to clear.

Materials

Per pair of students:

1. Two to four specimen tubes (2.5 cm × 7.5 cm) each containing mud with *Tubifex*.
2. A bench lamp
3. 100 g of Plasticine
4. A 30-cm ruler
5. A retort stand and clamp
6. A stop watch

Per class:

1. A weighing balance

Methods

Placing a bench light above or beside a tube, the response of the worms should be observed to the following stimuli:

(a) moving a book between tube and lamp (movement plus change in light intensity)
(b) switching light on or off (change in light intensity alone).

No response to these stimuli should be detectable. It will, however, be found that worms respond clearly to (c) a tap on the bench. The response to a soft tap is to pause from wriggling, and to a hard tap is to withdraw into the mud for several seconds before emerging to recommence wriggling. The response to vibration can now be used to investigate the following questions:

(a) Is the length of the pause in wriggling correlated with the intensity of the stimulus?
(b) Does the response habituate to stimulus repetition?

It is important in all exercises that students should try to standardize stimuli. The taps on the bench may be simply standardized by dropping a ball of Plasticine of known weight from a given height, indicated by a ruler held vertically in a clamp, on to the bench.

The response will probably have to be ranked according to some arbitrary scale, since two types of response are involved, pause in wriggling and withdrawal into the mud. A simple scoring system would be:

Nature of response	Score
Hesitation in wriggling	1
Distinct pause in wriggling	2
Pause plus partial withdrawal	3
Pause plus complete withdrawal	4

The intensity of the response will be found to be positively correlated with stimulus intensity. The response will also be found to habituate quite readily, e.g. a light tap at 15-second intervals will habituate the response in 12-20 taps using a habituation criterion of no response in 3 successive taps.

Time

Ten minutes as a demonstration; 35 minutes as a simple class exercise with systematic measurement of a few animals in one or two different situations; more than 60 minutes for open-ended investigation and discussion.

3.4. Wriggling rate and defaecation

Purpose

There is a tendency for students commencing the study of behaviour to believe that 'simple' animals are indeed simple. This exercise looks at three measures of behaviour: wriggling rate, time between defaecation bouts, and number of faecal pellets per bout. The initial assumption of the students may well be that all three are behavioural measures of metabolic activity and therefore will probably vary together. The fact that there is no such clear correlation provides a basis for discussion.

Preparation

A specimen tube needs to be filled with about 3 cm of *Tubifex* mud and then filled to the top with water. The top of the tube must then be covered over with a net of about 1-mm mesh size, held in with an elastic band. The tube should then be immersed in a 12-cm diameter crystallizing dish and turned so that the net covered end is facing downwards. The tube should then be held in a clamp with the lower end of the tube still in the water (figure 3.4.1). Two such tubes per

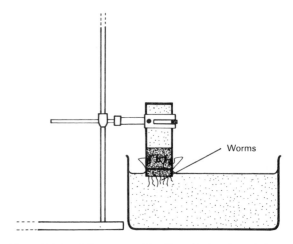

Figure 3.4.1 *Tubifex* worms with tails projecting through the net covering at the bottom end of the specimen tube and into the water held in the crystallizing dish. The lower part of the tube is filled with mud and the upper part with water.

pair of students should be set up 24 hours before the practical session is due to start.

Materials

Per pair of students:
1. Two specimen tubes (2.5 cm × 7.5 cm) each containing mud with *Tubifex*.
2. About 20 cm^2 of 1-mm mesh net
3. One 12-cm crystallizing dish
4. Two retort stands with clamps
5. One bench lamp
6. A stop watch
7. Two elastic bands

Methods

After the tubes have been hanging in water for 24 hours, a number of worms will have pushed their tails through the net and be wriggling them in the water. With a bench light behind the tube, it is possible to see the faecal pellets being extruded by individual animals and falling down through the water. The faeces break up into pellets of more or less uniform length, which allows them to be conveniently scored. Defaecation occurs in distinct bouts, during which the gut contents from most of the projecting part of the animal are voided in less than half a minute; there is then a pause of 1-10 minutes before the next bout. Several worms should be chosen for investigation, some having a fast wriggle rate and some a slow one. The following features should be scored for each individual:

(a) Time taken to perform 50 wriggles.
(b) The time from the *start* of one defaecation bout and the *start* of the next.
(c) The number of faecal pellets per bout.

The number of faecal pellets per bout will be found to be rather constant (8-12) even between individuals with very different wriggle rates. The time between the start of defaecation bouts is, however, very variable and not obviously correlated with wriggle rate. The problems of interpreting wriggle rate were raised in Section 3.1 but, even though it may be a measure of the animals' metabolic activity, it does not correlate with food intake as measured by time between defaecation bouts. However, the number of faecal pellets per bout remains more or less constant for all animals.

Time
60-90 minutes including discussion time.

3.5 Clustering or ball formation in *Tubifex*

Purpose
If a large number of *Tubifex* worms are scattered in a dish of tap water, they eventually clump together in a tight ball. The reasons for this are not immediately clear and therefore allow room for speculation. These balls of worms give some appearance of behaving like single organisms, which would of course necessitate 'communication' between worms. How much, therefore, can the observed behaviour be accounted for in terms of simple responses of individual organisms acting independently?

Preparation
One 7-cm diameter crystallizing dish per pair of students should have its floor covered with 1 cm of mud and be filled with water at least an hour before the start of the class to allow the water to clear.

Materials

Per pair of students:
1. A solid clump of *Tubifex* about 2 cm in diameter
2. Two 7-cm diameter crystallizing dishes
3. One 7-cm diameter crystallizing dish with the floor covered with 1 cm of mud and topped with water
4. A stop watch or clock

Methods
A crystallizing dish is filled with tap water, and a 1-cm diameter clump of *Tubifex* placed in it and shaken till the worms are scattered over the floor. The behaviour of the worms should then be observed. After a couple of minutes small clumps of worms have formed; when these contact neighbouring ones, they appear to reach out and draw one another together. This process continues until, after 8-10 minutes, only one or two tight balls remain. If, however, a 1-cm ball of worms is placed in the dish with the mud, after 10 minutes it will be seen that the worms are burrowing into the mud and the ball breaking up. It can therefore be brought out in discussion that what may appear at first glance to be cooperation between individual worms may be no more than the attempts by every individual worm to cover its body surface (thigmotaxis). When the worms are given the choice of covering their body surface with other individuals or with mud, they prefer the latter, if for no other reason than possibly because it also provides food.

Time
From 35 minutes, including time for discussion.

4. Sand Shrimps

INTRODUCTION

Many of the larger intertidal amphipods (Crustacea) are useful experimental animals in the behaviour laboratory. There are, however, two main problems. Foremost is the identification of the animals, and secondly there is the problem of supply. The simple studies outlined here are based on work with gammarids (*Marinogammarus* spp.) but similar and interesting studies can be conducted with, for example, the more robust but less easily handled sand hoppers (Talitridae) from the top of the shore, or the small carnivorous gallery-building amphipod *Corophium volutator* (Corophiidae), which can be sieved in large numbers from muddy sand in certain sheltered localities. The main groups of amphipods can be distinguished and separated fairly quickly with the aid of a text such as N. B. Eales (1961), *The Littoral Fauna of the British Isles*.

In the case of the gammarids, identification down to species can be a laborious task where large numbers of animals are involved. But in practice, we have found that at the level of study indicated here identification to species is not critical. The student can quickly check that all shrimps have antennules twice the length of their antennae and, if the animals have all been collected from the same microhabitat on the shore, it is likely that they are all of one species.

Large numbers of *Marinogammarus* can be collected with relative ease from under stones and weed on the middle and upper shore. The most useful piece of collecting gear is a small metal-rimmed sieve or net which can be scraped through the gravel, sand and debris to pick up the animals. The debris and animals are then knocked out of the sieve into a bucket. The gammarids will travel well in the damp gravel and weed. In the laboratory, the animals can be stored either in damp weed in covered pie dishes in a cool refrigerator, or kept in shallow trays of cold seawater. It is possible to maintain them for several weeks, but we have always found that the best results are from specimens which have been freshly obtained. Ideally, the animals should be used within three days of collection.

EXPERIMENTAL WORK

4.1. The 'escape' response

Purpose

When the tide is out, the sand shrimps are found under stones and weed in often surprisingly small films, pockets and shallow pools of residual water. The most noticeable feature of their behaviour when uncovered by the collector is their determined scurrying and swimming to fresh cover. This burst of activity, frequently referred to as an 'escape' behaviour by students, can be demonstrated simply under laboratory conditions, and may be an ideal starter for a more detailed analysis of the animal's response to light.

Preparation

The gammarids should be left quietly in their covered trays for at least 10 minutes prior to the demonstration.

Materials

Per demonstrator or per group of students if they are to follow up the demonstration immediately.
1. One shallow tray of seawater
2. 20-30 sand shrimps
3. Cardboard (or other opaque material) to cover tray
4. Bench lamp

Methods

The demonstration is very simple. The animals are in approximately 1 cm of seawater in a covered tray. The bench lamp is on, so that it will light up the contents of the tray when the cardboard cover is removed. The cover (the laboratory equivalent of the rock or weed on the seashore) is lifted and the gammarids burst into activity, swimming and scuttling about the tray. Replace the cover and leave it in place for at least 2 minutes. The demonstration of activity following removal of the cover can then be successfully repeated.

It is suggested that the following investigations then form a neat introductory pattern of simple behaviour investigations.

1. Is it vibration or the sudden increase in light which triggers the burst of activity? This can quickly be checked by allowing the animals to settle in a tray in dim lighting and then switching on a bench lamp. The sudden increase in light intensity produces the so called 'escape activity'. The sand shrimps similarly respond to tapping the tray and to water movement.
2. Follow up the response to a sudden increase in light intensity with exercises 4.2 and 4.3.
3. When students have worked with the animals in these simple introductory exercises, they will have encountered the problem of 'clumping', i.e. where the gammarids cluster together after a brief period of random movement, also males carrying females, and possibly also the response to contact with debris or the sides of the tray. This can lead naturally into the open-ended exercise outlined in 4.4.

Time

This introductory demonstration is usually done to set the scene for further investigation and, with discussion, should take no more than the first 10 minutes of a class.

4.2. An investigation of the sand shrimps' response to light

Purpose

To investigate more fully the burst of locomotory activity which follows a sudden increase in light intensity.

Preparation

None

Materials

Per group of 2-3 students:
1. Tray of seawater as in 4.1
2. 20-30 sand shrimps
3. Cardboard cover for tray
4. Bench lamp
5. Fine sand or silt

Methods

Agitate a little fine sand or silt in the tray of seawater containing the gammarids and allow it to settle. (Frequently the fine detritus and silt inevitably collected with the amphipods is sufficient.) The idea is to try to show up the tracks and disturbances made by the animals in this layer of fine debris. The student can now lift the cardboard cover after a while and note that the directions taken by the sand shrimps are apparently at random. A general moving about in all directions takes place until the animals settle, either in clumps or tucked into the edges and corners of the tray. This type of behaviour pattern has been termed an *orthokinesis* (Fraenkel and Gunn, 1961). In this case the animal moves more quickly in bright light, and more slowly and stops in darkness.

The student should then arrange things so that half his tray is in bright light and half in reduced light (e.g. the cardboard covering one half of the tray). The gammarids quickly swim and scurry about until the majority, if not all of them, come to rest in the shaded part of the tray. This redistribution of the animals in a half-shaded tray takes a matter of seconds rather than minutes, and thus is a conveniently speedy and reliable experiment. With a two or three-minute rest under the cover, the animals will react to the light and distribute themselves again, and with further rests will continue to do so for 5 or 6 trials. With the repeats and possible variations in students' approach, two points may arise:

(a) There is a time lag between the lamp going on and the beginning of the burst of swimming. This can be compared with a similar phenomenon in mosquito larvae (14.2).

(b) The apparent strength of the response seems to

weaken with each repeat, thus suggesting *habituation* to the stimulus. This is the basis of the next exercise (4.3).

Time
20-30 minutes.

4.3. Habituation of the response to a sudden increase in light intensity

Purpose
To study the weakening of the response to the repeated presentation of the light stimulus, and to demonstrate that it is in fact habituation rather than fatigue.

Preparation
The sand shrimps should be left undisturbed in their covered trays of seawater for 10-15 minutes prior to the tests.

Materials

1. Tray of seawater as in 4.1
2. 20-30 sand shrimps
3. Cardboard covers for tray
4. Bench lamp
5. Stop watch

Methods
This can either be done as a demonstration with a small class, or as individual or small-group exercises. The methods may vary in minor details at the discretion of the tutor or student group, but essentially the task is to present the amphipods with a bright light at a standard distance for a standard period of time. The period of rest should likewise be kept the same. Suggested values are:

(*a*) 40-watt bulb at 30 cm above tray,
(*b*) light switched on for 10 seconds,
(*c*) cover replaced or light switched off for 60 seconds.

With this timing regime habituation to an increase in light intensity should be obvious within four or five trials. The amount and duration of locomotion falls off noticeably after the second period of light.

The animals can then be left for a slightly longer rest – say, three minutes, and the response will be seen to have recovered.

To test the idea that it is fatigue rather than habituation to a specific stimulus, simply tap the tray or drop a few drops of seawater on to the surface of the water in the tray; the immediate burst of activity will convince the experimenter that it cannot be simply fatigue.

Time
20 minutes.

4.4 Contact responses

Purpose
Clumping of individuals and aggregations in corners of the tray and around pieces of detritus during previous experiments should leave the student in no doubt as to the importance of contact responses to this intertidal animal. This exercise is designed to lead the student into a longer and possibly open-ended investigation of locomotion, leg function and contact responses.

Preparation
None

Materials

Per student or small group of students:
1. Tray of seawater as in 4.1
2. 20-30 sand shrimps, including some males carrying females
3. Cover for tray
4. Bench lamp
5. Fine watchmaker's forceps
6. 2 toothpicks
7. Small beaker containing small pebbles, pieces of shell and seaweed
8. Access to supply of boiling water
9. Dissecting instruments
10. Stereomicroscope
11. Hand lens
12. Plasticine

Methods
The sand shrimps should be kept quietly in the shallow tray of seawater, shielded from direct light by the cardboard cover. On lifting the cover, the now familiar response to light occurs. As the shrimps move about, a small piece of shell or a pebble (1 cm across) can be placed in the tray and the reactions of the gammarids noted. The animals usually bump into the

obstacle and stop moving with their bodies held closely to it. If examined closely with a hand lens, it will be noted that the gammarids are holding on to the piece of shell with their longer thoracic limbs. These thoracic limbs 6, 7 and 8 are held reflexed back and out from the body, and are used when the animal 'walks' on its side and (in this case) holding or maintaining contact with its environment (figure 4.4.1).

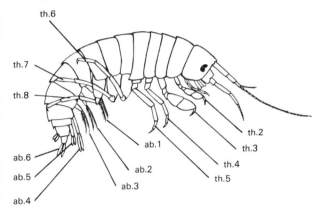

Figure 4.4.1 Gammarid showing limb types: th = thoracic, ab = abdominal

Students may wish to investigate the role of the thoracic legs 6, 7 and 8 further by trying to touch the uppermost two or three while the shrimp is moving on its side. This can be done with fine forceps, a splinter from a toothpick or piece of fine wire. Touching thoracic 8 alone frequently stops the animal, but with a little experimentation it will become clear that the shrimp requires thoracic 7 and 8 at minimum in contact with an object before it settles for any length of time. Students will also find that if thoracic 7 or 8 are held gently but firmly with the forceps the shrimp will stop passively. Holding any other appendage will produce violent curling and attempts to jerk free. These simple tests help to reinforce the idea that the reflexed thoracic legs play an important role in their contact responses.

If a shrimp which is maintaining contact with a piece of shell, or weed, is held firmly, with forceps, by the end of a free reflexed thoracic limb, it is possible to lift the animal and the piece of shell clear of the water. The small hooks or claws at the end of these longer thoracic limbs obviously give the animal a surprisingly firm hold on to the substrate.

An open-ended type of investigation is possible here. We suggest it takes initially one of the following lines.

1. *A general investigation into locomotion and leg functions.* After initial tests as suggested above the student should kill a gammarid by dropping it into boiling water, and then study the morphology of the different limbs. With further observations on the living animal the student should quickly establish that the first two pairs of legs are subchelate and are assisted by the subchelate third and fourth pairs which are turned forward when feeding. Thoracics 6, 7 and 8 have already been discussed. The abdominal limbs (pairs 1-3) can be seen to be involved in swimming, while the last three pairs (abdominal limbs 4, 5 and 6) are found to be largely concerned with kicking at the substrate when it begins to swim, or jerking forwards when hopping out of water (figure 4.4.1).

   The function of the subchelate thoracic limbs 2 and 3 can also be investigated in males holding females. Males holding females can be dropped into hot water to kill them. The male remains holding the female and the interesting asymmetric hold can be examined under a microscope. A male will take hold of a female which has been killed by crushing her head, but quickly drops her again. If the female is killed by dipping in hot water, the male shows no interest in her at all. (As a side line, the killed females can be examined to find eggs or young in the brood pouch.)

   The functions of abdominal limbs 1, 2 and 3 can also be further investigated. They are involved in producing respiratory currents. These can be demonstrated by putting Indian ink or milk into the water beside a resting animal.

2. The investigation could also initially concentrate on the responses of the sand shrimps to contact. Different sizes and shapes of Plasticine (or Perspex) objects can be presented to the moving shrimps, and their responses observed. The minimum size of object required to trigger the grasping and contact responses can be found.

   Coupled with this, a series of tests could be devised where different limbs and portions of limbs are removed with fine scissors to see how the contact responses are modified.

   It can be established fairly quickly that the shrimps do not have a preferred side, i.e. they do not tend to prefer walking or swimming on their right side. Their swimming, indeed, is frequently upside down and they may first make contact with an obstacle in that position.

   A further and final note – the depth of water is important in the experimental apparatus. The animals respond to the surface film. So, here again, is a possibility for further testing.

Time

30-60 minutes.

REFERENCES

Eales, N. B. (1961), *Littoral Fauna of the British Isles,* 3rd ed., Cambridge Univ. Press.
Fraenkel, G. S. and Gunn, D. L. (1961), *The Orientation of Animals,* New York: Dover.

5. Woodlice

INTRODUCTION

Woodlice are the most successful of the few terrestial crustacea. These small isopods are common and worldwide in occurrence. Virtually all must inhabit damp or humid microhabitats such as rotting vegetation, beneath stones or the bark of dead trees, where they are sheltered from desiccation. Woodlice lose water very quickly, and their ecology is dominated by the need to conserve water. In general they are scavengers, eating plant materials, living as well as decaying, and they may also take a small amount of animal matter, tiny arthropods and carrion. Breeding takes place usually in the warmer months, and they may produce three broods of young per year. The eggs and developing young are carried in a brood pouch for about five weeks.

In the British Isles, the two commonest genera are *Porcellio* and *Oniscus*. Both are easily found under stones, bark, mats of vegetation and similar sheltered places, throughout the country. They are slow-moving and completely harmless, and so are simply collected; they make particularly convenient experimental animals, equally for the young school pupil and the older student. Only two points require emphasis, the animals are delicate and should be handled with care, and the animals must always be kept in a moist environment at an even temperature. A useful general account of the biology and identification of British species may be found in *Woodlice* by Sutton (1972).

The housing for woodlice is fairly simple. A large glass trough with a cover to prevent drying out, and a number of layers of damp moss and pieces of bark or rotten wood is all that is required. This should be kept in a dark, cool place; not on the windowsill! Woodlice can live up to four years in captivity, if cool damp conditions are maintained.

EXPERIMENTAL WORK

5.1. The distribution of woodlice in a humidity gradient

Purpose

This, if properly set up, is a neat simple exercise in observation, recording and interpretation. The woodlice are introduced into a chamber where a humidity gradient has been set up. The student can observe the initial movements of the animals, and record and interpret the final distribution of the woodlice.

Preparation

Careful preparation of the humidity-gradient chamber is all-important. The plan of a possible chamber is shown in figure 5.1.1. Notice that the size of the chamber is important. Because woodlice show contact responses, i.e. their behaviour changes when they come into contact with, for example, the walls of the container or any irregularity in the substrate, it is essential that the floor or arena on which they are to move be as level and smooth as possible, and that the curve of the arena walls be as gentle as possible. Unsatisfactory experimental results become commoner

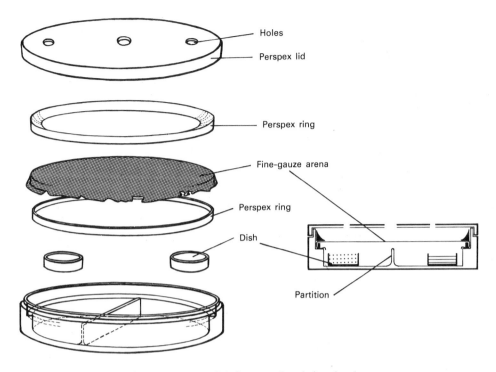

Figure 5.1.1 Exploded diagram of a choice chamber.

the smaller the humidity-gradient chamber becomes. The diameter of the arena should not be less than 30 cm. A large pie-dish or pneumatic trough is the smallest container that should be considered. The arena surface should be carefully prepared from the finest mesh materials. Perforated zinc sheet, expanded metal sheet, and various arena materials have been tested by the authors and, where the perforations are at 0.5 mm or more, results have deteriorated. The best results came from fine-mesh nylon net curtaining material carefully stretched on embroidery hoops which just fitted the inside diameter of the pneumatic trough used. A circle of Perspex, as shown in figure 5.1.1, forms a neat gently-curving wall to the arena to reduce the possibility of contact responses to a minimum. The water and desiccant must be in position in the chamber for one or two hours before the experiment is carried out. Finally, the chamber must be shielded from direct light. A cylinder of black paper round the chamber is usually adequate to prevent light responses interfering with the animals' responses to the humidity gradient.

Materials

Per student or group of students:
1. Humidity-gradient chamber as described and illustrated in figure 5.1.1, set up as discussed at least an hour beforehand
2. Access to a vivarium containing woodlice
3. 10 woodlice. (These should be left undisturbed in the vivarium until the experiment is about to start.)
4. Small beaker for holding woodlice immediately prior to experiment
5. Small paintbrush for brushing woodlice into containers
6. Strip of smooth paper to be used as a chute for introducing the woodlice into the experimental chamber
7. Stop clock

Methods

The woodlice should be collected from the vivarium where they have been kept. Handle the animals as little as possible. Where possible, brush them gently into the small beaker off the pieces of moss and bark from the vivarium. Discard any inactive, moribund or damaged specimens. Uncover the hole in the centre of the chamber lid and, with the aid of the paper chute, slide the ten woodlice on to the centre of the fine-mesh

c

arena. Cover the hole in the lid again. The woodlice normally scatter randomly across the arena, and the majority end up moving round the edge of the chamber. The students can then record the positions of the animals on blank plans of the chamber at five-minute intervals. The majority of the woodlice should be found in the humid region of the chamber after a short time, and significant results can be obtained well within 40 minutes.

If poor results are obtained check the following:

(a) The junction of the wall and the arena floor, a small hole, snag or unevenness can cause the animal to rest there.
(b) The animals themselves may be already in poor condition or damaged.
(c) Incident light may be interfering with the experiment.

Time
The class time is approximately 40 minutes. The preparation time is about five minutes per chamber, plus settling time of at least an hour, so that the humidity gradient can establish itself.

5.2 Turn correcting in the woodlouse

Purpose
In 1961, Dingle demonstrated that, if a box elder bug *Leptocoris trivittatus* was forced to make a right-angle turn in the corridor of a maze, it tended to turn in the opposite direction when emerging from the maze; this was called *correcting behaviour*. Hughes (1966 and 1967) demonstrated the same correcting behaviour in *Porcellio scaber*. This response is now known for a number of arthropod species, e.g. the mealworm *Tenebrio molitor* (Dingle, 1964). This suggests that if you cannot get *Porcellio* it would be interesting to try any small arthropod that will run readily through mazes.

This orientation response is easy to elicit and is influenced by a number of variables which can easily be manipulated in simple experiments. It is therefore an ideal system for an investigation of the control of an orientation response.

Preparation
The mazes used are not yet available commercially,

so will have to be made in the workshop. It is possible to make more simple mazes than the ones described here, but they are robust, easy to use in the laboratory, and permit a flexible experimental situation. If the mazes have been made, no further preparation is needed to the start of the class.

The mazes should be cut from 13-mm thick Perspex 25 mm broad; the corridors for the woodlice to run along should be 8 mm or 9 mm wide for woodlice, and cut about 7 mm into the Perspex. A set of mazes should consist of:

(1) Two straight mazes 5 cm long
(2) One straight maze 10 cm long
(3) One 90° angle maze with long arm of 10 cm and short arm of 5 cm (figure 5.2.1.)
(4) One 60° angle maze with long arm of 10 cm and short arm of 5 cm
(5) One 30° angle maze with long arm of 10 cm and short arm of 5 cm

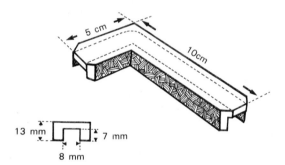

Figure 5.2.1 A 90° Perspex maze. Detail shows dimensions of an arm of the maze in section.

Materials

Per pair of students:
1. Ten woodlice
2. One set of mazes (as above)
3. One pair of compasses (for drawing a circle)
4. One protractor
5. One stop clock or watch
6. Two 7-cm diameter crystallizing dishes (to contain woodlice)
7. One bench lamp

Per class:

1. Some leaf litter
2. A small pot of black enamel paint
3. A very fine paint brush

Methods

Effect of angle of forced turn

A circle of radius 5 cm should be drawn on a piece of paper and the 90° maze placed on the paper (corridor down) with the end of the short arm of the maze at the centre of the circle. A pencil line should then be drawn round the outside of the maze. A bench lamp should be placed with the bulb about 30 cm immediately above the maze exit.

A woodlouse is then taken from the 'before test' crystallizing dish and run through the maze, and the point at which it crosses the circle is marked. The woodlouse is then placed in the 'after test' crystallizing dish. All ten woodlice should be run through the maze at least twice. It is very important that both crystallizing dishes should contain damp leaf litter to keep the woodlice humid between tests, or they will die.

Having run the woodlice through the 90° maze, they should then be similarly run through the 60°, 30° and a 15 cm straight maze (made by placing a straight 5 cm and a straight 10 cm together).

The results may then be plotted as the mean or median angle of correction against the angle of turn of the maze. (Angles turned in the same direction as the forced turn are scored as negative angles.)

It will be found that there is a positive correlation between angle of forced turn and angle of correction.

Effect of distance walked before and after the forced turn

Once the correcting behaviour has been demonstrated, the question arises as to how the direction preference is established and how it is extinguished. It could be that the distance walked before the turn established the strength of the direction preference, and the distance walked after the turn and before the choice weakened the correction response. With the apparatus provided it is possible to vary the length of 'in' and 'out' arms.

Hughes (1967) showed that distance run before the turn did not affect the response, and our observations in the class situation agree with this. (It is interesting that the woodlouse is apparently unlike the mealworm in this respect, since the latter was shown by Dingle (1964) to be influenced by the distance before the turn.)

It is possible to show that as distance between turn and choice increases, so the angle of correction decreases, i.e. confirming Hughes' (1967) observation (figure 5.2.2).

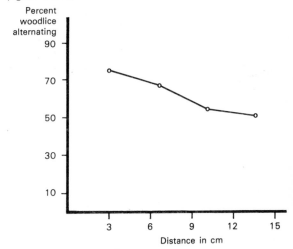

Figure 5.2.2 Percentages of woodlice *Porcellio scaber* turn correcting at 4 different distances from a 90° forced turn (from Hughes, 1967).

The absence of any effect of distance travelled before the forced turn on correction response suggests that the turn itself initiates the response. Hughes (1967) suggested that the greater distance walked by the legs on the 'outside' of the woodlouse during the turn caused the effect. This was ingeniously confirmed by Beale and Webster (1971) using a straight runway with one side a moving walkway moving against the direction of movement of the woodlouse. The technically minded may wish to make such a maze for class use, but we have not used one and gather that they are inclined to swallow up the woodlice in the mechanism!

Effect of detention in the maze after forced turn

The fact that the 'memory' of the forced turn appears to decrease with increased walking after the turn could be due not to the distance walked but to the time between forced turn and choice. This can be tested by holding the woodlouse in the end of the maze before allowing it to choose its direction. Hughes (1967) was able to show that elapse of time alone caused a diminution of the response, but in the class situation we have found this to be not very obvious because of erratic behaviour of the woodlouse after being allowed out of the maze.

Influence of antennae on turn correction

A point which is noticed by students in the practical class and which they will follow up in an open-ended class situation is the fact that there is often a big scatter in the correction angle. Careful observation suggests that if the woodlouse emerges from the maze slightly asymmetrically with the antenna of one side only touching the maze, the woodlouse turns strongly towards the side of contact just as it loses contact with the maze. This suggests a thigmotactic response mediated by antennal stimulation. If the antennae are snipped off, the scatter of the correction angle can be seen to be reduced.

Influence of light on turn correction

Quite simply, paint over the compound eyes of woodlice and this will be found to have no obvious effect on mean correction angle or scatter; i.e. providing the experimental situation is uniformly lit. If, in addition to the overhead bench lamp, there is appreciable directional light coming from a window, students may notice that, if the woodlice emerge from the maze *towards* the window, the scatter of the correction angle is increased as compared with animals which emerge from a maze pointing *away* from the light. This exercise can therefore be used to emphasize the use of proper controls.

Effect of desiccation on turning

Hughes (1967) observed that woodlice showed a stronger turn correction when the maze was illuminated by a bench lamp overhead than when there were room lights only. He suggested that this might be due to the maze being warmer and therefore more desiccating with the bench lamp. We have not tested this, but it should be simple in a class situation to keep one lot of woodlice (controls) in a beaker with damp leaf litter for about 10 minutes and another (desiccated) in a beaker without leaf litter, and then test for correction angle in the same maze. Hughes (1967)

points out that the improved correction for 'desiccated' animals may not be the direct effect of the desiccation but due to the fact that desiccated animals run faster than non-desiccated ones (Gunn, 1937) and that this reduces the *time* between forced turn and choice, and that causes the improved correction. This suggests another project: that of testing for a correlation between running speed and turn correction, which emphasizes the flexibility of this exercise where simple apparatus makes possible a whole range of experiments.

Time

To investigate systematically the effect of the angle of forced turn and of the distance walked before and after the turn requires 3 hours. If a more open-ended approach is adopted, where the students are instructed to observe the effect of running the animals through a 90° maze and then to test their own hypotheses from then on, it is still possible to get useful results in 3 hours, provided that students are encouraged to concentrate on one or two experiments rather than dabble. The exercise is particularly suited to a more extended project. It can provide a good 30 hours' work spread over two or three weeks.

REFERENCES

Beale, I. L. and Webster, D. M. (1971), 'The Reference of Leg Movement Cues to Turn Alternation in Woodlice (*Porcellio scaber*)', *Anim. Behav.*, 19: 353-356.

Dingle, H. (1961), 'Correcting Behaviour in Boxelder Bugs', *Ecology*, 42: 207-211.

Dingle, H. (1964), 'Correcting Behaviour of Mealworms (*Tenebrio*) and the Rejection of a Previous Hypothesis', *Anim. Behav.*, 12: 137-139.

Gunn, D. L. (1937), 'The Humidity Reactions of the Woodlouse (*Porcellio scaber*)', *J. exp. Biol.*, 14: 178-186.

Hughes, R. N. (1966), 'Some Observations on Correcting Behaviour in Woodlice (*Porcellio scaber*)', *Anim. Behav.*, 14: 319.

Hughes, R. N. (1967), 'Turn Alternation in Woodlice (*Porcellio scaber*)', *Anim. Behav.*, 15: 282-286.

Sutton, S. L. (1972), *Woodlice*, London, Ginn & Co.

6. Shore Crabs. Comparative Studies

INTRODUCTION

Of the dozen or so crabs common intertidally in British waters, we suggest that *Carcinus maenas, Cancer pagurus* and *Portunus puber* are particularly suitable for simple comparative-behaviour studies. *Carcinus maenas,* the common shore crab, is a particularly hardy shore species, ranging from the pools high on the shore down to the *Laminaria* zone and beyond. It is found on a wide variety of substrate: rock, weed, gravel, sand or mud are apparently equally acceptable. Similarly, this species appears to be largely indifferent to the degree of salinity or exposure. It is at home on an exposed headland or in a sheltered estuary. This makes the common shore crab a readily-available robust laboratory animal.

Cancer pagurus, the edible crab, is less robust in some respects than the common shore crab, being less tolerant of lowered salinities and lack of suitable cover. The rock and boulder shores washed by the full-salinity waters of the open sea are the typical habitat of this species. Wherever there are such conditions with plenty of crevices and loose rock for cover, the edible crab is common. But its range on the shore is more restricted than that of the common shore crab; in general it tends to be restricted to regions near low water mark, though very young specimens can be found higher up the shore. Like *C. maenas, Cancer pagurus* is widely distributed and common round the coasts of the British Isles but, whereas the common shore crab is an all-the-year-round member of the intertidal fauna, *Cancer* is a seasonal visitor, being common in the summer months and then moving offshore in the winter. The vast majority of specimens of *Cancer* collected intertidally

will be juveniles. The larger specimens tend to be at extreme-low-water spring-tide levels and below.

Portunus puber, the Fiddler or Velvet swimming crab is a most handsome species with its fine covering of velvety hair, clean blue and orange joints, blue lines on its legs, and bright red eyes. It is quicker moving, more active, and perhaps more belligerent in disposition than the other two species. Like *Cancer* it is a clean clear full-salinity water species, and it is common on the lower shore under weed and stones. It is, however, a more southern species than either *Carcinus* or *Cancer,* and tends to be restricted to the south and west coasts of the British Isles.

All three crabs are easily collected by turning over rocks and weed on the lower shore. *Cancer* is perhaps the most easily lifted by hand, the other two require care on account of the nippers, particularly in the case of large specimens. The most aggressive crab can be lifted safely by holding it with the thumb on one edge and the index or middle finger on the other edge of the carapace, so that the legs and claws are pointing away from the hand. Some students prefer to have a piece of wood, a ruler, or the handle of a net which they use to pin down the crab until they can get a firm grip of the two edges of the carapace. Certainly with *Portunus* this is a safer method for the beginner. The same handling procedure should be used in the laboratory.

The three crabs are readily distinguishable and can be identified easily with the use of a text such as Collins' *Pocket Guide to the Seashore* (Barrett and Yonge, 1972). The colour variation in *Carcinus* can mislead the beginner, but will be found extremely useful in the laboratory, where it will make identification of individuals that much easier. The identification of individuals of *Portunus* and *Cancer* is best done by marking the top or edge of the carapace carefully with a small file. The colour variations in these two crabs are rarely sufficiently marked to allow easy recognition of individuals. The sex of mature crabs is easily established. The shape of the abdomen is diagnostic; if narrow and pointed the crab is a male, if larger and rounded it will be a female (figure 6.0.1).

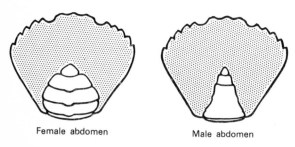

Female abdomen   Male abdomen

Figure 6.0.1 Ventral view of carapace of *Carcinus maenas* to show the difference in abdomen shape between mature males and females.

The crabs are best transported in damp weed in a wooden box or sack. It is important to keep the crabs apart as far as possible, so that fighting and the resulting damage and loss of limbs is reduced to a minimum. The more weed cover they have during transport the less fighting there is. In the laboratory the crabs can be stored in a number of ways. Ideally they should be kept in a temperature-controlled aquarium room at about 10°C. But where this is not available, they can be kept in small containers with a large surface area and about 5-7 cm seawater, i.e. enough to cover the animals, in a cool fridge or a cool room out of direct light. Feeding is easy; the crabs will accept almost anything meaty, e.g. white worm *Enchytraeus,* earthworm, dog food, liver, etc., but it is vitally important that excess food is removed, so that the seawater remains sweet. Constant changing of the seawater helps if this is possible. Otherwise, an aerator or bubbler will be found useful, and adequate for short periods.

EXPERIMENTAL WORK

6.1. Burying in sand

Purpose

The three species of crab are found with partially overlapping distributions on the seashore. This overlap could lead to specialization of the types of retreat favoured by the three in order to avoid competition. The partial separation of the species may equally have led to some specialization in the selection of retreats due to differences in habitat. All three species are likely to find in their habitats mud, sand or gravel; this offers a potential retreat from tide or predator. The problem therefore is whether each species is capable of exploiting it.

Preparation

The crabs should be brought into laboratory aquaria at least 4 days before they are to be used, to allow them to acclimatize to the increase in temperature. If an aquarium room is used, the temperature of which is somewhat lower than that of the class laboratory, then the temperature of water at the start of the practical exercise should be the same as the aquarium room although it will warm up, gradually, during the class.

Enough sand (preferably sea sand) should be collected to provide for the class. It is preferable to wash all sand, whatever the source, in seawater before use, to remove fine mud which will cloud the water in the class and make observation difficult.

Materials

Per 4 students:
1. Two *Carcinus maenas* (5-9 cm carapace width) (figure 6.1.1)
2. Two *Cancer pagurus* (5-9 cm carapace width)
3. Two *Portunus puber* (5-9 cm carapace width)
4. One plastic washing-up basin (30-40 cm across and about 15 cm deep), the floor covered with sand to a depth of 6-7 cm, and a depth of about 5 cm of seawater above that.
5. A bench lamp
6. A stop clock

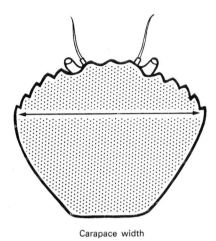

Carapace width

Figure 6.1.1 Dorsal view of a crab carapace to show measurement of carapace width.

Methods

One crab should be placed in the basin of sand and its behaviour observed and recorded in the form of detailed notes. After the crab has stopped burrowing, or after 3 minutes, whichever is the less, the crab should be placed back in its bucket and another crab tested.

After the two individuals of each of the 3 species have all been observed they should be tested again, this time using the stop clock to measure:

    (a) latency from entry into the basin to first burrowing movements
    (b) time from start to completion of burrowing.

After each crab has completed any burrowing that it is going to do, it should have sand pushed over it so as to bury it to a depth of 1 or 2 cm. It should then be observed for another 3-4 minutes to record its response.

The following results should be obtained:

Carcinus maenas

This species burrows very readily and quite rapidly into the sand. There are some individual differences in the digging pattern, and it is possible that there are variations with size which would be interesting for someone to look at but, for crabs in the 7-9 cm carapace width range, there are certain easily recognizable and apparently stereotyped behaviour components.

    (a) *Clawing.* The walking legs of both sides move as though the crab is trying to walk in both directions at once; this results in the crab digging itself down into the sand.

    (b) *Pushing away.* The two chelipeds are moved from the front of the body outwards, pushing sand away from the front of the crab.
    (c) *Rocking.* After the body of the animal has begun to sink into the sand, it begins to rock back and forth, thereby digging itself further into the sand.
    (d) *Jerk.* One or more vigorous downward jerks of the body, occurring when burying is quite far advanced. The most obvious effect of it is to cause sand to flow over the top of the carapace helping to conceal the crab.
    (e) *Shuffle.* Not always occurring but, when present, seen after the jerks. The movement is a series of rapid, slightly rotating, oscillations which cause sand particles to flow over the carapace, sometimes almost completely covering it.
    (f) *Knees up.* Shown by a minority of crabs and not observed to occur when shuffle is present. It is seen after the jerks and takes the form of the last pair of walking legs being flexed alternately over the top of the carapace thereby spreading sand over it.

In *Carcinus*. therefore, burying behaviour is well developed. Burying starts quickly after the crab has contacted the sand and is complete within about 30 seconds; it results in the carapace being at least partially and sometimes completely covered with sand.

When artificially covered with sand to a depth of 1 or 2 cm, *Carcinus* gradually tilts the body until the eyes, antennae and antennules are just protruding, and the exhalant water current is unobstructed. This response occurs usually within 1-2 minutes.

Cancer pagurus

Three of the same components of the burying behaviour occur as in *Carcinus;* these are clawing, pushing away, and rocking. Their appearance in comparison with *Carcinus* is, however, rather laboured and clumsy. No components of burying are seen which are not present in *Carcinus.* The sequence of behaviour components is therefore like a slowed-down attenuated version of that seen in *Carcinus.* The result of the burying in *Cancer* is also less effective as the whole of the top of the carapace remains visible above the sand, although the legs are well dug in. The burying latency is rather longer than in *Carcinus,* but the duration of the attempted burying is about the same.

If a *Cancer,* having completed burying, is covered with 1 or 2 cm of sand, it does not attempt to alter its position to any extent. The disturbance of the sand indicates that it has started respiratory water currents again, and the antennae may stick up through the sand; but it does not seem keen to raise itself up so that antennules and eyes protrude. If buried under 3 or 4 cm of sand, the crab may make no apparent move in 10 minutes.

Portunus puber

This species shows a reluctance to bury in sand at all and, when this is attempted, it is with the back of the crab against the side wall of the basin. The burying behaviour, when it occurs, forms an interesting comparison with the other species. Only three components of burying behaviour occur, and these are the same ones as are shown by *Cancer* – clawing, pushing away and rocking; the latter is a rather exaggerated tilting back and forth of the body. The movements are, however, surprisingly ineffectual, as the crab remains almost completely exposed except for the legs, which become buried in the sand. The delay from start of test to burying, when it occurs, is usually more than a minute, and the duration of the burying attempt is shorter than in *Carcinus*.

If *Portunus* is covered with a 2-cm layer of sand, it quickly digs itself upwards and lifts the whole of the front of the carapace free from the sand, so that it is clearly visible.

The conclusion of this exercise seems to be that *Carcinus* is well adapted to using sand as a retreat, and prefers to rest just below the sand with eyes, antennae and antennules monitoring events above sand level. *Cancer* is inefficient at burying and perhaps relies to some extent on its heavy armour to protect it from predation. If covered by drifting sand it does not, however, hurry to dig itself out as it can easily do. *Portunus,* on the other hand, is not only reluctant to bury itself but, when buried, quickly digs itself out. Although a comparatively lightly armoured crab, it does not seem to use sand as a refuge, but probably relies more on speed and agility to evade predators (see Exercises 6.5 and 6.6).

Time

About 1½ hours to observe and record all the situations systematically. Additional time for a discussion of the similarities and differences between species is advisable.

6.2. Crawling under rocks

Purpose

These three species of crab are likely to find in their habitat crevices under and between rocks which are potentially places of concealment and protection in addition to burying in the sand as investigated in 6.1. This exercise is a comparison of the response of the three species to rocks and rock crevices.

Preparation

Crabs should be adapted to laboratory temperatures in advance of the class as suggested in Preparation 6.1.

Enough rocks or pieces of brick should be gathered to provide at least two rocks of a size that could easily conceal a crab, for each pair of students. If these are not taken from the seashore, they should be soaked in seawater for 24 hours.

Sufficient sand should also be collected to provide a 1-cm floor covering for all the test bowls.

Materials

Per 4 students:

1. Three plastic buckets, ⅓ full of seawater. (Each bucket should contain 4 crabs of one species only; these are for the use of 2 pairs of students.)

Per pair of students:

1. Two *Carcinus maenas* (5-9 cm carapace width)
2. Two *Cancer pagurus* (5-9 cm carapace width)
3. Two *Portunus puber* (5-9 cm carapace width)
4. One plastic washing-up basin (30-40 cm across and about 15 cm deep), the floor covered with a 1-cm depth of sand.
5. Two or three rocks or pieces of brick 15-20 cm across
6. A bench lamp
7. A stop clock

Methods

The bench lamp should be positioned to shine into the basin, and one rock should be laid on the sand so that there is no cavity under it. Each crab should be placed separately into the basin, and its behaviour observed and recorded over a 3-4 minute period. The second rock should then be placed in the basin, and the two rocks arranged so that a cavity is formed under them, sufficiently large for a crab to crawl into. All the crabs should then be tested again.

It will be observed that if there is no cavity, *Carcinus* moves to the side of the rock and then tries to dig itself into the sand; if there is a cavity, the crab very readily crawls into it, although poking into the cavity with a pencil usually persuades the crab to leave it.

Cancer also backs up to a rock and presses itself against it if no cavity is present. It also very readily crawls into the cavity if one is provided. The behaviour of *Cancer* once in a cavity is, however, markedly different from that of *Carcinus;* it extends its legs and presses itself strongly against the rock walls. It cannot be teased out of the cavity with a pencil. By getting hold of the crab and trying to pull it out, it can be appreciated how tightly wedged the crab is; this is a perfectly safe thing to do, because the crab in this position makes no attempt to nip, the claws being held passively in front of the body, presumably as protection.

Portunus differs from the other two species, chiefly because of its reluctance to use the rocks as cover. If there is no cavity under the rock, *Portunus* will continue to walk around or stand in the open. When a cavity is provided, the crab approaches it with caution and may not fully enter it for 3 or 4 minutes. Once in the cavity, the crab seems ready to stay there. If disturbed with a pencil, the crab is reluctant to come out; it may nip at the pencil, but does not press itself against the side of the cavity like *Cancer*. We have observed a *Portunus* entering a cavity and then proceeding to excavate the sand from it in a most elaborate way. The crab walked sideways dragging the trailing claw and front leg like a dredge bucket; outside the cavity the sand was dispersed with long pushing-away movements of both claws. This kind of behaviour will probably not be seen in the rather rushed class situation, but does serve to illustrate that, although *Portunus* is more reluctant to enter strange cavities, it is adapted to use them possibly on a long-term basis.

All three species, therefore, seem to show some adaptation to use rock cavities as retreats. It is known for the lobster *Homarus vulgaris,* at least on an anecdotal level, that particular individuals may occupy the same cavity for a long time. The same may also be true of these crab species.

Time

About $1\frac{1}{2}$ hours will be required if each crab is to be tested twice and full notes made. Additional time for discussion of the similarities and differences between species is also desirable.

6.3. Dark/light choice test

Purpose

In Section 6.2 it was observed that *Carcinus* and *Cancer* readily entered cavities under rocks, whereas *Portunus* showed some reluctance. A simple testable hypothesis to explain these observations is that *Carcinus* and *Cancer* have a strong preference for the dark, and *Portunus* a slight one. This exercise is a simple one to test that hypothesis. It therefore shows how observation in a semi-natural situation can lead to a rigorously designed experimental situation. In addition it illustrates how an experiment may be designed to control for possible confounding variables.

Preparation

Crabs should be adapted to laboratory temperatures in advance of the class, as suggested in Preparation 6.1.

Materials

Per 4 students:

1. Three plastic buckets, $\frac{1}{4}$ full of seawater. (Each bucket should contain 4 crabs of one species only; these are for use by 2 pairs of students.)

Per pair of students:

1. Two *Carcinus maenas* (5-9 cm carapace width)
2. Two *Cancer pagurus* (5-9 cm carapace width)
3. Two *Portunus puber* (5-9 cm carapace width)
4. An opaque white plastic tank approximately 45 cm × 25 cm × 12 cm deep, filled 7-8 cm deep with seawater
5. One piece of black plastic or Perspex to fit inside one end of the tank, i.e. about 22 ɑm × 12 cm
6. One bench lamp

Methods

The tank should be set up with the black panel in position at one end of the tank and the bench lamp over the mid-point of the tank. One crab should be taken from its bucket and placed in the mid-point of the tank facing '12 o'clock' (the ends of the tank being '3 o'clock' and '9 o'clock'). It should be held in this position for a count of three, to allow it to observe the difference between the two ends of the tank, and then released. The first end touched with at least one leg is scored as the chosen end.

This basic plan should be modified to incorporate controls for two possible confounding variables. The first of these is that a particular individual crab or species of crab may have a *handedness*, i.e. may have an internally determined preference to walk to the right rather than the left, or vice versa. This can be controlled simply by facing the crab on the first trial towards '12 o'clock' so that dark is to its left, and on the second trial towards '6 o'clock' so that dark is to its right (figure 6.3.1).

The second variable is that of asymmetry of the choice situation, other than that produced by the black test surface; this could be produced by the presence of a window, or unknown factors outside the tank. The control for this is to place the dark panel at the other end of the tank and test the crab again (figure 6.3.2.) Each crab should therefore be tested in four situations:

(*a*) Facing 12 o'clock; black 3 o'clock
(*b*) Facing 6 o'clock; black 3 o'clock
(*c*) Facing 12 o'clock; black 9 o'clock
(*d*) Facing 6 o'clock; black 9 o'clock

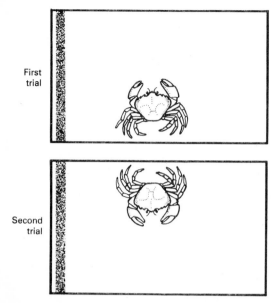

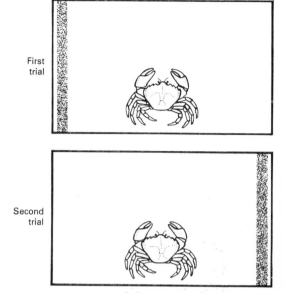

Figure 6.3.1 Control for 'handedness' in crabs. The first dark/light choice is given to a crab facing 12 o'clock and the second to it facing 6 o'clock. (The size of the crab is exaggerated for clarity.)

Figure 6.3.2 Control for external stimulus bias. The first dark/light choice is given with the dark surface on the left and the second with it on the right. (The size of the crab is exaggerated for clarity.)

At lower teaching levels, this procedure may be given on a worksheet, together with the reasons for it. At higher levels the possible presence of these confounding variables may be pointed out, and students asked to devise their own controls.

The result of the experiments is broadly to confirm the hypothesis based on the observations of Section 6.2. *Carcinus* makes a rapid and very consistent choice for the end with the black panel, rarely choosing the pale one. *Cancer* is always delayed in making its choice due to the adoption of a 'cataleptic' posture each time it is handled (figure 6.3.3); this will last for

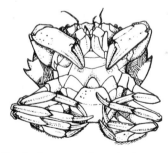

Figure 6.3.3 Cataleptic posture of a crab with legs and claws rigid and curled under the carapace. Most readily shown by *Cancer*, but shown here in *Carcinus* (after Jensen, 1972).

half a minute or so, after which the crab makes a highly consistent choice for the end with the black panel. The preference in *Portunus* for the black panel is not so marked as in the other species; more than one-third of all choices may be to the pale end. Since this result is consistent with the observations of Section 6.2, it demonstrates that the response of crawling towards a cavity under a rock is explicable as a simple visual response to low light intensity. It does not however, rule out the possibility that other factors might be exercising some effect on the choice.

There is one additional feature of the experiment which is worth mentioning, and this concerns the antennules (figure 6.3.4). In crabs the antennules are conspicuous for their little flicking movements. In *Portunus* they are somewhat longer than in *Carcinus*, and in *Cancer* shortest of all, but in all three species they are quite easily seen, and in all three species they show little flicks. The antennules are capable of being pointed both to the front, the left or the right, or even one right and one left.

Before moving, a crab placed in the choice box can be seen to test the water to the left and to the right with its flicking antennules. Just before the crab moves, its decision is indicated by both the antennules being faced in the direction of walking (figure 6.3.4). They are then held in this position while walking continues.

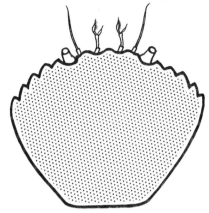

Figure 6.3.4 Dorsal view of carapace of *Carcinus maenas* to show position of antennae and antennules. The latter are shown pointed to the right to 'taste' water currents coming from the right-hand side.

There is good evidence that these antennules bear numerous chemoreceptors, so this antennule movement shows that the crab is looking for chemical cues in addition to visual ones before deciding to move in a particular direction. This does not explain the purpose of the curious flicking of the antennules. A study by Snow (1973) on the hermit crab *Pagurus alaskensis* suggests that the reason for this is that there is a tuft of chemo-sensory hairs at the end of each antennule, positioned in such a way that during a flick they are forced apart by the water flow, thereby greatly increasing the sensory surface area (figure 6.3.5). The crab is therefore sampling the water in a series of chemosensory pulses.

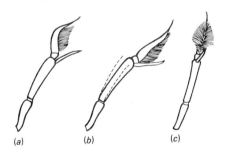

Figure 6.3.5 A single antennule of *Pagurus* to show: (*a*) resting position; (*b*) (side view) and (*c*) (front view). Movement of antennule during a flick to shown terminal bunch of sensory hairs forced apart during a flick by water resistance (after Snow, 1973).

Time
It will take at least 1½ hours to give each of the six crabs its 4 choices and get all the information recorded. Additional time for discussion is desirable.

6.4. Left and right 'handed' walking preferences

Purpose
When humans start to walk, they have to decide whether the right or left foot goes first; a crab has to decide whether to go to the left or the right. When all other things are equal, rather than be rendered immobile with indecision, a crab might toss a mental coin (i.e. go randomly left or right) or have a built-in preference (e.g. if in doubt, go left). A built-in preference might be characteristic for individuals or possibly for a whole species. This exercise is designed to test for such handedness, using the apparatus of Section 6.3 slightly modified.

Preparation
Crabs should be adapted to laboratory temperatures in advance of the class, as suggested in Preparation 6.1. If it is planned to look systematically at the effect of a missing claw on left/right preference, then the claws should be broken off (some right and some left) at least 24 hours before the crabs are to be tested.

Materials

Per 4 students:

1. Three plastic buckets, ⅓ full of seawater. One for each species of crab, as in 6.1.

Per pair of students:

1. Two *Carcinus maenas* (5-9cm carapace width)
2. Two *Cancer pagurus* (5-9 cm carapace width)
3. Two *Portunus puber* (5-9 cm carapace width)
4. An opaque white plastic tank, about 45 cm × 25 cm × 12 cm deep, filled 7-8 cm deep with seawater.
5. Two pieces of black plastic or Perspex (about 22 cm × 12 cm) to fit in the ends of the tank)
6. One bench lamp

Methods
The tank should be set up with the bench lamp over the mid-point and a black panel in position at each end. It is preferable to have both ends black rather

than both ends white, because it appears to persuade animals to make up their minds faster, a dark end being more attractive than the exposed middle. Each crab should be placed in turn in the middle of the tank, facing 12 o'clock and, after it has made a choice, it should be tested again facing 6 o'clock to control for possible asymmetry of the choice situation.

The result of the experiments is to show that none of the species has a clear directional preference. Individual animals seem to have slight preferences, and it would be interesting to know if these persist over longer periods of time.

Animals with missing claws

It is very likely that among the crabs in the class are one or two with missing claws, and additional clawless crabs should be prepared if required. This does not affect their ability to walk, but it could affect the preferred direction of walking. The results of the choice of these animals may therefore be compared with those of intact animals. It will probably be found that in *Portunus* the absence of a claw makes it more likely that the crab will walk in the direction corresponding to the side of the missing claw (clawless side first). In *Cancer* there is no obvious effect, and in *Carcinus* there is some indication that the crab prefers to walk with the clawless side last (i.e. opposite to *Portunus*). It is not easy to rationalize such a result. All that can be said is that it is interesting to observe that there are differences between the species.

Time
To give each of the six crabs four choices and record the results will take at least 1½ hours. Additional time for discussion is desirable to speculate on similarities and differences between the species.

6.5. Locomotion in the water and on land

Purpose
One of the obvious anatomical differences between the three species is that the last pair of walking legs in *Cancer* are quite pointed, in *Carcinus* slightly flattened, and in *Portunus* quite flattened and paddleshaped (figure 6.5.1); this suggests some difference in locomotion. The habitat difference which makes *Carcinus* more likely to find itself out of water than the other two might cause it to be more effective at walking on land. This exercise is designed to investigate the locomotory abilities of the three species.

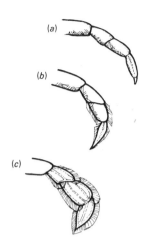

Figure 6.5.1 Terminal segments of the last walking leg of the 3 species of crab described to show modification for swimming: (*a*) *Cancer* – no flattening of leg and no fringe of hairs (*b*) *Carcinus* – leg slightly flattened and with a fringe of hairs (*c*) *Portunus* – leg very flattened and paddle-shaped and with a marked fringe of hairs.

Preparation
Crabs should be adapted to laboratory temperatures in advance of the class as suggested in Preparation 6.1. Sea sand is desirable (but not essential) and should therefore be collected and washed before the class if required.

Materials

Per 4 students

1. Three plastic buckets ⅓ full of seawater, one for each species of crab, as in 6.1.

Per pair of students:

1. Two *Carcinus maenas* (5-9 cm carapace width)
2. Two *Cancer pagurus* (5-9 cm carapace width)
3. Two *Portunus puber* (5-9 cm carapace width)
4. One plastic washing-up basin (30-40 cm across), half-filled with seawater and the floor covered with a 1-cm depth of sand.
5. One plastic bucket full of seawater
6. A blunt wooden or wire rod 20-25 cm long
7. A bench lamp
8. A stop clock

Methods

In the water
Each crab should be placed, in turn, in the basin and allowed a minute or so to settle down. If it walks about spontaneously, then the nature of the walking should be observ-

ed and recorded; whether fast or slow, with the body slightly or much raised above the substrate, etc. After the crab has been left for about a minute to walk about, it may be encouraged to move by tapping it gently on its legs or on the top of its carapace with a rod.

It will be seen that the undisturbed locomotion of *Cancer* is an even gliding movement with the body slightly raised from the substrate, but not fast. When poked with a rod, its speed is not much increased and if severely disturbed, it adopts the cataleptic position. *Carcinus* has an even gliding motion with the body hardly raised above the substrate; when disturbed it is capable of a quite rapid gliding run. The undisturbed locomotion of *Portunus* resembles closely that of *Carcinus* but, if the legs of one side are touched, the animal does a sudden vigorous jump to the other side. It is difficult to see how this is achieved, but the hind legs are certainly involved in the jump and the glide which follows. If very disturbed, *Portunus* shows vigorous beating of the legs which briefly lift it off the substrate.

Falling through water
The ability of a crab to swim or at least slow down its rate of descent through water can be tested by taking a crab and holding it just submerged in a full bucket of water for a count of 5 and then letting go. If this is done, it will be seen that *Cancer* adopts the cataleptic position when held and, when released, falls face forwards on the floor of the bucket with a heavy thud. If held till its legs are extended and then dropped, it still falls heavily and does not beat its legs during the descent. *Carcinus* extends all legs prior to being released; as it falls it beats its legs, maintaining a correct body position and probably also slowing up the fall. In *Portunus,* the legs, including the modified back legs, beat during the descent, slowing up the rate of fall.

Walking on land
If each crab is placed on the bench top and its walking observed, it will be found that *Carcinus* walks fairly readily with its body held well above the ground and the claws pointing forwards. *Cancer* moves slowly and clumsily, with body held low and bumping on the ground, and the claws held into the body. *Portunus,* though it may move quite rapidly, is frequently reluctant to move and drags its body along the ground. None of the species, therefore, seems well adapted to walk on land but *Carcinus* is, as expected, better than the other two.

These observations on the locomotion of the three species of crab reinforce the picture built up in Sections 6.1 and 6.2. *Cancer* is a slow-moving crab relying on heavy armour for protection. *Carcinus* is more flexible; it walks quite fast when submerged; when on land, it burrows readily and hides in rock crevices. *Portunus* is a fairly lightly-armoured crab, like *Carcinus,* but is able to move quickly and even swim over short distances.

Time
It will take about 1 hour to test all six animals once in each of the 3 situations.

6.6. Defensive and protective behaviour

Purpose
One of the more painfully obvious features of crabs is that they nip or bite with their chelipeds. This presumably provides some benefit to the crab in its natural habitat. This exercise uses the same simple situation as Exercise 6.5 to investigate the nipping behaviour and other protective devices.

Preparation
Crabs should be adapted to laboratory temperatures in advance of the class as suggested in Preparation 6.1. Sea sand is desirable but not essential, and should therefore be collected and washed before the class if required.

Materials

Per 4 students:

1. Three plastic buckets $\frac{1}{3}$ full of seawater, one for each species of crab, as in 6.1.

Per pair of students:

1. Two *Carcinus maenas* (5-9 cm carapace width)
2. Two *Cancer pagurus* (5-9 cm carapace width)
3. Two *Portunus puber* (5-9 cm carapace width)
4. One plastic washing-up basin (30-40 cm across), half-filled with seawater and the floor covered with a 1-cm depth of sand.
5. Two 25-cm lengths of balsa wood rod, one about 0.5 cm and the other 1.0 cm square section.
6. A dissection blade
7. A bench lamp

Methods

Strength of nip
Each crab in turn should be lifted out of the bucket and have the smaller (0.5 cm) balsa rod pushed into its claw until it grasps it. When the crab has released the rod, the part of the rod with the indentation should be cut off with a blade and marked to indicate what species caused the bite. After at least two nips have been recorded, the strength of the nip can be compared between the species by comparing the indentations in the balsa wood. (If this is going to be done on a more systematic basis, an arbitrary scale of strength may be used.) *Carcinus* and *Portunus* are about the same strength, neither making much impression on the rod. *Portunus* will be found rather more reluctant to nip than *Carcinus*. *Cancer* will be found to nip even when the legs are in the immobile cataleptic position (figure 6.3.3). The strength of its nip is dramatically more powerful than in the other species. A 9-cm carapace-width individual can severely maul even the larger (1.0 cm) balsa rod.

Other protective devices

It has been mentioned in Exercises 6.3 and 6.5 that when lifted up *Cancer* adopts a quite characteristic immobile cataleptic posture (figure 6.3.3). The claws are held downwards and open, and all the legs are curled up under the body. It is not at all obvious what advantage this may be to the crab, since it appears then to be both vulnerable and docile, except when objects are placed directly in the claws. One must only presume that somehow it is advantageous.

If *Portunus* is poked with the balsa rod from the front, it will be seen that usually the crab makes no attempt to nip but shows a rapid 'handing off' movement rather like a rugby player handing off an assailant. At the same time as handing off to one side the crab moves swiftly to the other.

Sometimes when disturbed in the water *Portunus* will suddenly tilt its body and extends its claws right out to the side. The effect is not only to increase suddenly the apparent size of the crab but to reveal the pale facial and inner claw surfaces, heightened by the orange red integument of the cheliped joints and the blue spots on the chelae themselves. It is not easy to persuade *Portunus* to show this defensive display on land, but *Carcinus* will show it in a slightly less dramatic form both on land and, to a lesser extent, in water (Jensen, 1972) (figure 6.6.1). This same display does also occur in *Cancer,* but here the claws are held more forward in a 'readiness' rather than a 'display' position — and it is anyway not easily elicited.

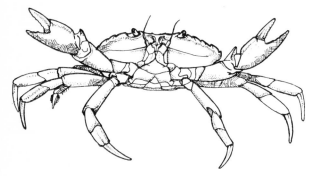

Figure 6.6.1 *Carcinus* in defensive display showing claws extended (after Jensen, 1972).

Time

It will take about 1 hour to test all six animals in water and on land.

6.7. Feeding and food detection

Purpose

The chelipeds of crabs are not only of defensive function as shown in Section 6.6 but also potentially of use in offence and in food manipulation. Before feeding, crabs must of course detect and locate food. This exercise briefly looks at detection and location of food and feeding behaviour.

Preparation

Crabs should be adapted to laboratory temperctures in advance of the class, as suggested in Preparation 6.1. Crabs should be at least 24 hours starved prior to the practical, but it is advisable anyway not to feed crabs at all when they are being held in aquaria. This prevents the water going 'sour' and the crabs survive well for at least two weeks.

Materials

Per 4 students:

1. Three plastic buckets $\frac{1}{4}$ full of seawater, one for each species of crab, as in 6.1.

Per pair of students:

1. Two *Carcinus maenas* (5-9 cm carapace width)
2. Two *Cancer pagurus* (5-9 cm carapace width)
3. Two *Portunus puber* (5-9 cm carapace width)
4. One plastic washing-up basin (30-40 cm across) half-filled with seawater and the floor covered with a 1-cm depth of sand
5. One plastic bucket full of seawater
6. A 200-ml beaker containing several pieces of crab food (bivalve flesh, e.g. mussel *Mytilus edulis,* fish, *Tubifex,* bits of meat)
7. A bench lamp

Methods

Firstly the two *Carcinus* crabs should be placed in the basin and allowed a minute or so to settle down before a piece of food is dropped in, not too close to either individual. After a delay of a further minute, during which very little appears to happen, the antennule activity of the crabs starts to increase with more moving back and forth and a higher flicking rate. There may then begin to be an opening and closing of the mouthparts, opening and closing of the claws, and placing empty claws to the mouth. The crab may then begin to move about in a not obviously directed manner. If the two crabs meet one another, they do not treat one another as food; a leg may be gently tested with a cheliped and then released.

Crabs may not notice the position of the piece of food, even when very close to it, although they are clearly excited by its proximity; however, as soon as a leg touches the food, the crab makes a very swift directional response to the food, which may be rather clumsy but shows that there are chemoreceptors in the legs of the crab which tell the crab that food has been touched and which leg has located it. The food is then grasped in one or both claws, stuffed into the mouthparts and chewed.

This pattern of behaviour is almost identical for *Cancer* and *Portunus,* which suggests that these species have very poor ability to locate food visually, and their distance chemoreception also lacks a directional capability. The latter conclusion is, however, probably unfair because of the artificial situation of the plastic basin. Water currents produced by the crab are going round and round the basin destroying any olfactory gradients. In a more natural habitat crabs may be able to detect water currents and walk 'upstream' to a food source. This is difficult to test in a simple experimental situation.

Time
35-40 minutes.

REFERENCES

Barrett, J. and Yonge, C. M. (1958), *Pocket Guide to the Seashore* (Collins).
Jensen, K. (1972), 'On the Antagonistic Behaviour in *Carcinus maenas* (L.) (Decapoda)', *Ophelia,* 10: 57-61.
Snow, P. J. (1973), 'The Antennular Activities of the Hermit Crab *Pagurus alaskensis* (Benedict), *J. exp. biol.,* 58:745-766.

7. Hermit Crabs

INTRODUCTION

Pagurus bernhardus is the only really common hermit crab whose juvenile forms may be collected intertidally around the British coasts. The adults are common from the sublittoral fringe into deep water, and the usual supply for laboratory work is collected by dredge. In the south and west *P. prideauxi*, a somewhat smaller relative of the above species, is common offshore but is rarely encountered in any numbers intertidally. Likewise *P. cuanensis* is almost invariably sublittoral.

Juvenile or adult, the hermit crab makes a fascinating subject for the behaviour laboratory. The animal's large and unprotected abdomen is normally inserted into a suitably sized empty gastropod shell. The modified uropods grip the shell, so that the hermit crab can drag this ready made 'house' along with it and retreat into it when danger threatens. This well-known aspect of its biology has been studied in detail, and much of the behaviour associated with the 'house' is readily reproducible in the laboratory.

The juveniles can be collected by hand from the lower levels of the shore. They can be particularly abundant in spring and early summer on shores where there are patches of shelly sand lying in pools of residual water. The young hermits select empty periwinkles *Littorina*, dog whelks *Nucella*, and topshells *Gibulla* as their early 'houses'. The young collector may frequently spot a periwinkle moving at a most unperiwinkle-like speed across the floor of a pool, and on capturing this 'mollusc' is surprised and delighted to find a young hermit crab in the shell.

The adults, however, must be ordered through a supplier of such marine specimens. It is most important that correct aquarium conditions are set up for hermit crabs. The seawater in the tanks must be clear, clean, well oxygenated, and kept at an evenly maintained low temperature (e.g. 10°C) out of direct lighting. The success of much of the behaviour work will depend on the animals being in good condition prior to and during the laboratory work. This is particularly true of the adults. The smaller juveniles seem to be a little more tolerant. They can be kept in smaller containers in a cool refrigerator for a reasonably long time. In all cases the seawater must be kept clean, and care taken with feeding. The hermit crabs will feed readily on a wide variety of flesh. Crushed periwinkles, white worm *Enchytraeus*, dog foods and other meats are accepted.

The most convenient size of crab to use has a shell aperture diameter of about 1 cm.

EXPERIMENTAL WORK

7.1. Feeding

Purpose
To demonstrate the location of food from a distance and the use of claws and legs in feeding.

Preparation
Although young hermit crabs seem to adapt quite readily to changes in temperature, it is preferable to bring them in to room temperature two or three days before they are to be used, to allow them to settle down.

Materials

Per pair of students:

1. Two hermit crabs in an 18-cm diameter crystallizing dish half-filled with seawater and the floor covered with clean beach sand.
2. One live mussel *Mytilus edulis*

Methods

The crabs should first be left undisturbed in their crystallizing dish to allow the respiratory currents of the crabs to set up a fairly regular movement of water about the dish. The water current flowing towards a crab, which may be seen by the flow of fine particles suspended in the water, is sampled by the flicking movements of the antennules whose hairs are chemosensory (Snow, 1973). If an opened mussel shell is placed in the water a few centimetres from either crab, after a short delay the rate of antennule flicks will be seen to increase, and the crab may make small claw and leg movements. Shortly after this the crab will begin to move actively about the dish, often moving quite directly towards the food source.

When the crab reaches the mussel, it tears at the flesh vigorously with one or both claws. The shell is held steady by the legs and sometimes one claw. The lumps of food are stuffed into the mouthparts by the claws and slowly chewed.

Time

10 minutes

7.2. Threat behaviour

Purpose

Juvenile hermit crabs are often found on seashores in moderately high densities, so that not only must they frequently come in contact with one another but they must also be in competition for food and for gastropod shells. This competition has resulted in the evolution in hermit crabs of stereotyped threat postures, to which subordinate animals respond by retreating. This exercise investigates threat behaviour and the response of crabs to threat.

Preparation

Crabs to be used for the class should be brought in to room temperature two or three days before they are to be used.

E

If crab skeletons are to be used for models, these will need to be prepared carefully in advance. The skeleton of a freshly killed crab should be dismembered, have the flesh removed, and the leg and claw joints stuck together again.

Materials

Per pair of students:

1. 4-6 crabs in an 18-cm diameter crystallizing dish half-filled with seawater and the floor covered with clean beach sand.
2. Two crystallizing dishes 12 cm diameter with 5 cm depth of seawater and the floor covered with sand
3. A glass rod, about 20 cm long
4. A lump of Plasticine
5. *Either* – The parts of a complete hermit crab skeleton: cephalothorax, legs and claws, *or* a piece of stiff, dark-coloured card about 10 cm square.

Methods

Initially the crabs should be watched as they move undisturbed about the large crystallizing dish. Notes should be made of any behaviour which looks like that of a 'dominant' animal. To students unfamiliar with animal behaviour, the subordinate behaviour will be the easier to recognize and is of two basic forms: retreat and withdrawal into shell. Both of these are usually seen in response to the approach and threat of a larger crab.

Threats are not so immediately obvious to an inexperienced observer: firstly, because they may not be readily distinguished from normal locomotion or posture and, secondly, because a threat posture may not be responded to by withdrawal of the threatened crab, and may actually induce reciprocal threat. Once recognized, however, threats will be seen as fairly stereotyped and of three basic types widespread among hermit crab species (Hazlett and Bossert, 1965):

(a) Major claw presentation (figure 7.2.1)
(b) Major claw extension (figure 7.2.2)
(c) Single or double leg raise (figure 7.2.3)

To observe these threat and subordinate behaviour patterns in a more simple situation, two crabs of approximately equal size should be placed together in a 12-cm diameter crystallizing dish and the behaviour observed.

The important visual components in the threat

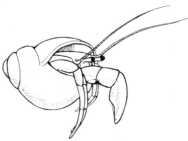

Figure 7.2.1 Major claw presentation in which the largest of the two claws is held extended towards the opponent in an upright position (after Hazlett, 1965).

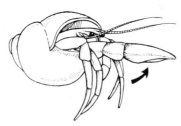

Figure 7.2.2 Major claw extension in which the largest claw is quickly raised towards the opponent (after Hazlett, 1965).

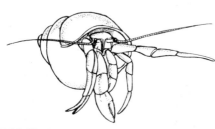

Figure 7.2.3 Single leg raise seen from the position of the opponent. The crab holds the leg raised to the side (after Hazlett, 1965).

posture may be investigated in a more systematic way by making models of a crab by mounting parts of a crab skeleton on to a ball of Plasticine to simulate normal position, the various threat positions, a model with no claws, and one which is just a piece of Plasticine. These models may then be presented on the end of a glass rod to a single crab in a dish and its response observed. Simpler models may be tried with pieces of card stuck into the Plasticine.

In each case such aspects of the response of the crab should be measured as: retreat, threat, retreat with threat, withdrawal into shell, and no response. The results may not be very clear cut, but pooling class results may manage to show significance for some stimulus features. Hazlett (1968, 1969) shows

that in *Pagurus bernhardus,* claw presentation, claw extension and *double leg raise* all produced more retreats than normal body position. He also showed that a model raised up on its toes, or with one or two extra large chelipeds produced more retreats than normal; however, rather curiously he discovered that a model with its large cheliped on the left side, i.e. opposite to normal, was less effective in inducing retreats than a normal crab.

Time

The observational part of the exercise should take about 20 minutes, and a brief experimental investigation using models could be accomplished in a further 30 minutes; however, a systematic experimental investigation is suitable for a project study.

7.3. Entry of a naked crab into a shell

Purpose

The ability of hermit crabs to identify and enter the spiral shells of gastropod molluscs is a remarkable adaptation. The anatomical specializations of hermit crabs can be readily seen in a naked individual when compared with an unspecialized crab such as the shore crab *Carcinus.* As well as these anatomical specializations, there are behavioural ones which may be studied in this exercise.

Preparation

Crabs should be brought in to room temperatures two or three days before they are due to be used.

Materials

Per pair of students:

1. 4-6 crabs in an 18-cm diameter crystallizing dish half-filled with seawater and the floor covered with clean beach sand
2. One crystallizing dish 12 cm in diameter with 4 cm of seawater and the floor covered with ½-1 cm depth of clean beach sand

Methods

First a crab must be removed from its shell. This is not a particularly easy business, as no method works very quickly or reliably.

One method which always fails is to try to pull the crab from its shell; the abdomen of the crab grips so strongly to the inside of the shell that the animal is torn in half before it will let go.

Another method sometimes suggested is to heat the tip of the shell with a match or soldering iron; apart from being unsuitable for large classes, many crabs are prepared to boil in their shells rather than come out.

The method we would recommend is to hold the shell in one hand and tap it rapidly and quite vigorously with a blunt hard instrument, such as the handle of a pair of scissors. The shell may be held with the palm of the hand upwards so that the crab may crawl out onto the fingers, but a slightly more effective method is to hold the shell in the water so that the crab can crawl out directly onto the substrate. It usually takes two or three minutes of tapping before the crab emerges, and even this method is not infallible, some crabs absolutely refusing to leave their shells.

Entry into an empty shell: A naked crab should be placed in a 12-cm-diameter crystallizing dish, together with the shell from which it was removed, and its behaviour observed. If the shell is orientated with its opening facing upwards, the crab will briefly place its large claw into the shell aperture, quickly slip its abdomen in, turn the shell over and walk off. If the shell is facing downwards, the crab will turn the shell over and enter the shell as described. The shell can therefore be recognized by the crab from various angles and treated accordingly.

Entry into a shell full of sand: If a crab is presented with a shell filled up with sand and its aperture facing upwards, the crab will first push its cheliped into the sand. This seems to tell it not only the size of the shell cavity but also whether the shell is completely filled with sand – because, if the sand does not fill the shell cavity, the crab may push its abdomen through the sand and right into the shell. However, if the shell is completely filled with sand, crabs will quite often rotate the shell to remove the sand and appear to be able to tell which direction of rotation is the more effective. This would repay further investigation.

Once in the shell, the hermit crab exhibits two easily recognizable behaviour patterns to remove sand from the shell: *pumping* and *shell tipping*. In the first, the crab makes repeated, sudden jerks back into the shell forcing out water and a shower of sand. In the shell tipping, the crab stands firmly on the substrate and, by rotating its abdomen, tips the shell up on end so that sand pours out.

Entry into an occupied shell: If a naked hermit crab is given a shell of approximately the size it requires, already occupied by a smaller crab, it will fight with the shell occupant, often evicting it. The fight starts with the naked crab climbing on to the occupied shell and turning it over; it then grabs hold of the occupant by the claws or legs, and may try to push its small claw into the shell cavity. The resident may resist passively for several minutes, but then suddenly capitulates and leaves the shell, half climbing out and half pulled, neither crab being in the least bit damaged in the process.

Time
40-50 minutes to observe all three situations suggested.

7.4. Changing shells

Purpose
Hermit crabs change to larger and larger gastropod shells as they grow. This suggests that they select shells that are neither so small that they may not be retreated into, nor so large that the crab has to carry around excess weight. We might therefore expect a crab on encountering a better shell to move from its own into it. This may be studied in this exercise.

As hermit crabs seem to live in situations where there is a scarcity of suitable shells, a crab must often find, on locating a better shell, that it is already occupied. Such a crab will often fight the shell owner for the shell in a characteristic and ritualized manner, which may result in the two crabs changing shells. Shell fighting behaviour may be observed in this exercise.

Preparation
None

Materials
Per pair of students:

1. 6 crabs of a range of sizes in an 18-cm-

diameter crystallizing dish, half filled with seawater and the floor covered with clean beach sand

2. One crystallizing dish 12 cm in diameter with 5 cm depth of seawater and floor covered with clean beach sand

3. Four empty gastropod shells of a range of sizes approximately corresponding to those occupied by the crabs provided

Methods

Two or more crabs should be removed from their shells as suggested in 7.3 (Methods) and then allowed to enter shells which are either too large or too small; this they will readily do.

Changing to an empty shell: A crab which has been transferred to a shell either too large or too small should be placed in a dish with an empty shell of about the same size as that from which it was originally removed, and the behaviour of the crab observed. The crab will first grasp the empty shell in its legs, apparently to assess its size (Reese, 1963). If the shell seems suitable, the crab turns the shell so that the aperture faces upwards, and inserts its large cheliped into the shell to assess the cavity for size or for obstructions. Finding the inside of the shell satisfactory, it will quickly slip out of its unfavourable shell into the better one.

If a crab is provided with a shell no better than the one it occupies, it may investigate without transferring. Even a crab left in its original shell, but provided with empty shells of a similar size, will investigate them without transferring; this suggests that even an optimal shell does not completely suppress the behaviour of investigating empty shells (Reese, 1963).

Changing to an occupied shell: Two crabs of rather different size should be transferred to each other's shells and placed together in a crystallizing dish. The larger will then approach the smaller and, after threatening it, will grab hold of the shell and turn it over; the larger crab will then seize the smaller by a leg or claw as described in 7.3 (Methods). If the size difference between crabs is great, then the smaller may rapidly release hold of its shell and come out; however, if the crabs are fairly closely matched, the dominant may start to show a different and ritualized form of behaviour. It will begin to hit the shell of the subordinate crab with its own in a rapid series of four or five taps which are easily audible. This shell tapping may be repeated at intervals for several minutes but usually results in the threatened crab leaving its shell.

Time

30-35 minutes to observe both situations suggested.

REFERENCES

Hazlett, B. A. and Bossert, W. H. (1965), 'A Statistical Analysis of the Aggressive Communications Systems of Some Hermit Crabs', *Anim. Behav.,* 13: 357-373.
Hazlett, B. A. (1968), 'Communicatory Effect of Body Position in *Pagurus bernhardus* (Decapoda, Anomura)', *Crustaceana,* 14: 210-214.
Hazlett, B. A. (1969), 'Further Investigations of the Cheliped Presentation Display in *Pagurus bernhardus* (Decapoda, Anomura)', *Crustaceana,* 17: 31-34.
Reese, E. S. (1963), 'The Behavioural Mechanisms Underlying Shell Selection by Hermit Crabs', *Behaviour,* 21:78-126.
Snow, P. J. (1973), 'The Antennular Activities of the Hermit Crab *Pagurus alaskensis* (Benedict)', *J. exp. Biol.,* 58: 745-766.

8. Cockroaches

INTRODUCTION

Cockroaches belong to the order Dictyoptera which includes both the cockroaches and the mantids. The cockroaches are the more unspecialized members of the order and do not show, for example, the modification of the first pair of legs characteristic of the mantids.

They are hemimetabolous insects, with only the adults bearing wings. The wings are not usually effective in flight. The forewings, which are somewhat more sclerotized than the hind wings, are sometimes referred to as the *tegmina*.

Cockroaches tend to be regarded as repulsive animals which live on the dirt and refuse associated with poorly maintained human habitation. This way of life, however, is only shown by a handful of species which, because of their close association with man have been carried all over the world. There are about 3500 species of cockroach, mainly of tropical distribution, ranging in habitat from desert to rain-forest.

The so-called American cockroach *Periplaneta americana* probably originated from Africa (Cornwell, 1968). It is now a world-wide domestic pest in kitchens and warehouses, and in outside habitats such as rubbish tips and sewers where the climate is sufficiently mild.

The sexes differ — the female has a ventral keel on the posterior end of the abdomen with a slit running along it. Both sexes have anal cerci but only the males have a pair of anal styles (figure 8.0.1). There are 6 nymphal stages. *P. americana* tends to be rather smelly and, to some, unpleasant to handle. It is also a particularly rapidly moving cockroach, which makes work with it difficult.

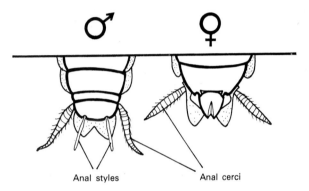

Figure 8.0.1 Ventral view of posterior segments of adult *Periplaneta americana* showing differences between male (left) and female (right).

A much more pleasant odour-free easy-to-handle cockroach is *Nauphoeta cinerea*. This is a mainly tropical domestic cockroach but worldwide in distribution. It is quite easy to breed, from egg to adult taking about 3 months. The nymphs may emerge from a single ootheca. When fully grown they are 2.5-3 cm in length.

A third species of use to the behaviour tutor is *Gromphadorhina portentosa* which originates from Madagascar. The adults have no trace of wings, but the males can be distinguished from the females by a pair of prominent tubercles on the pronotum (figure 8.0.2). The females are viviparous. The size when fully grown is about 7 cm. It is a large, pleasant, amiable cockroach; it is slow and ponderous in its movements, and therefore easy to handle; it also has no odour. One slight drawback is that it tends to be a rather slow breeder.

Figure 8.0.2 Adult *Gromphadorhina*, showing the complete absence of wings. The male (left) shows prominent tubercles on the pronotum, absent in females (right). Both animals are about 7 cm long.

Cockroaches can be maintained easily in a glass tank, e.g. small aquarium, fitted with a tight escape-proof cover. A smear of Vaseline around the top inside margin of the tank is sufficient to keep cockroaches off the cover, and therefore makes access to the tank easier. A 40-watt bulb is sufficient to provide heat. The bottom of the tank or jar should be covered in sawdust, and some empty eggboxes may be included to provide cover. An ample supply of water should always be available. There are a number of ways of doing this, but all devices should be proofed against animals falling in and drowning. We tend to favour wide dishes with a mound of water-saturated cotton wool. Food is little problem: starch wastes, bread, oatmeal, cereals, occasional greens and bacon fat are frequently used. Finally, it is important that cockroach tanks are cleaned and sterilized at regular intervals.

EXPERIMENTAL WORK

8.1. Observations on antennal grooming behaviour

Purpose
If the cockroach *Periplaneta* is observed in a home cage situation, individuals will be seen from time to time placing one antenna in the mouth parts and running it through from near the base to the tip (figure 8.1.1). This is an intricate and well co-ordinated piece of behaviour, involving movements of antenna, head and legs, illustrating behaviour involved with care of the body surface. The normal occurrence of this behaviour is infrequent, but can be simply and reliably reproduced under laboratory conditions. It is therefore both an interesting and convenient subject for study.

Figure 8.1.1 *Periplaneta* shown drawing its antenna through the mouthparts in a grooming movement.

Preparation
Before the class, cockroaches should be housed individually in transparent plastic sandwich-box-type containers (15 cm × 10 cm × 10 cm) with a coarse net roof covering, held on with an elastic band.

Materials

Per pair of students:

1. Two cockroaches in separate boxes
2. A small stoppered bottle containing a few ml of chloroform or dilute acetic acid
3. A fine paint brush
4. Some Vaseline (petroleum jelly)

Methods
At the most introductory level, antennal grooming by *Periplaneta* may be very quickly demonstrated to a class. The paint brush should be dipped in the chloroform and the tip applied to one of the antennae about half-way along. This provokes almost immediate antennal grooming.

The same procedure may be used as a class exercise, each pair of students provided with two cockroaches. In classes unused to handling insects the net should be kept on the box, and the tip of the paint brush introduced through one of the holes. For more-experienced classes, Vaseline should be smeared liberally round the inside rim of the container to prevent the cockroach from escaping, and then the paint brush may be freely applied to the cockroach's antenna and the behaviour observed in detail. The effect of the stimulation of the antenna will be to elicit antennal grooming which will be seen in *Periplaneta* to consist of the following components:

(*a*) Strong downward flexion of the stimulated antenna
(*b*) Downward bending of the head
(*c*) A slight rotation of the head to face *away* from the stimulated side
(*d*) Catching of the antenna in the angle of the tibia and tarsus of the *contralateral* prothoracic leg and placing the antenna in the mouthparts
(*e*) Returning the leg to the normal standing position
(*f*) Drawing the antenna forwards through the constantly moving mouthparts, forming an increasingly larger loop till the tip of the antenna is finally released (figure 8.1.1)

The point of detail which should be particularly noticed is that the stimulated antenna is placed in the mouthparts by the *contralateral* foreleg. This was also observed by Luco and Aranda (1964) for the cockroach *Blatta orientalis*; however, both *Gromphadorhina portentosa* and *Nauphoeta cinerea* will be found to use the *ipsilateral* foreleg.

Another important feature of this grooming behaviour is that, even within a single individual the response may be somewhat variable; for example, on occasion both prothoracic legs may be raised simultaneously to place the antenna in the mouthparts. The repetition of antennal stimulation several times on the two cockroaches provided should allow students to describe the main features of the behaviour and notice its variability.

These observations may then be used as a basis for discussion of the functions of the antennae, and the value of keeping body surface and particularly sense organs clean. The observations also raise questions about the control of the behaviour by the nervous system. If this exercise is carried out in conjunction with appropriate lectures or a dissection of the nervous system, then students may be led to appreciate

that perception of the stimulus applied and movement of the antenna must be respectively through sensory and motor nerves passing to and from the supra oesophageal ganglion; also movement of the mouthparts is controlled by motor nerves originating from the suboesophageal ganglion, and leg movement is initiated by motor nerves originating in the prothoracic ganglion (figure 8.1.2).

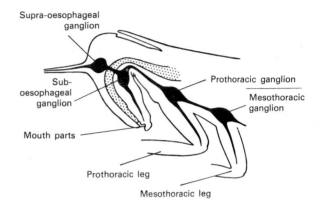

Figure 8.1.2 Side view of head and anterior thorax of generalized cockroach to show the positions of the ganglia relative to antennae, mouthparts and legs.

Variations in the appearance of grooming from one occasion to the next suggests that antennal stimulation does not simply release a fixed sequence of behaviour acts; however, it is not possible to tell from these observations whether sensory feedback from the leg placing the antenna or from the mouthparts receiving the antenna regulates the grooming sequence.

Time
20 minutes to 1 hour to complete systematic observations.

8.2. Development of alternative antennal grooming patterns after leg amputation

Purpose
It was observed by Luco and Aranda (1964) that if both prothoracic legs were amputated from the cockroach *Blatta orientalis,* which normally places its antenna in its mouthparts with the contralateral prothoracic leg, for the first 7 days or so after the

operation the animals used the *homo*lateral *meso*thoracic leg, but later changed to the contralateral one. This implies some nervous plasticity which allows a new response to appear. This could be due to a trial-and-error learning process, since the new response emerges after repeated ineffective or inefficient attempts; however, some developmental changes may also be involved. The acquisition of this new response may be studied using any convenient cockroach species, but is illustrated here by *Periplaneta* which uses the contralateral leg and *Gromphadorhina* which uses the ipsilateral one.

Preparation
Cockroaches should be individually housed in transparent sandwich boxes (15 cm × 10 cm × 10 cm) covered in net secured with an elastic band.

Materials

Per pair of students:

1. 10-20 cockroaches
2. A small bottle of chloroform
3. A fine paint brush
4. A stop watch
5. A tube of rapid drying glue

Per class
1. A CO_2 cylinder

Methods
Before an operation a cockroach should be anaesthetized with CO_2 from a cylinder. The part of the leg to be removed may be simply snipped off with a pair of scissors. Many different types of operation may be tried; for example:

1. Amputation of the tarsus from one prothoracic leg

2. Amputation of one prothoracic leg at the coxa-femur joint

3. Amputation of both prothoracic legs at the coxa-femur joint

4. Glueing one prothoracic leg to the thorax

Controls should be run in parallel with any operated group. These should be anaesthetized and given some kind of operation trauma, such as making a small cut in the abdomen away from the ventral nerve cord.

At least two (and preferably more) animals should be set up in each group and their grooming behaviour in response to chloroform stimulation observed as in Exercise 8.1. Observation should be carried out daily for 10-15 days.

Intact *Periplaneta* place the antenna in the mouthparts with the contralateral leg and *Gromphadorhina* the ipsilateral one. Both species are essentially unaffected by the removal of the tarsus from the leg required for grooming. This suggests that the tibial spines may be important in preventing the escape of the antenna, although they may have other functions.

After removal of the contralateral prothoracic leg, *Periplaneta* will be found to change immediately to the use of the ipsilateral mesothoracic leg. In the case of *Gromphadorhina* the removal of the ipsilateral prothoracic leg results at first in unsuccessful attempts to use the ipsilateral mesothoracic leg. These attempts improve slightly over the next few days.

On the day following the removal of both prothoracic legs, *Periplaneta* shows antennal flexion, and downward bending of the head. Neither mesothoracic leg is used at this stage to place the antenna, which in consequence frequently slips from under the thorax without reaching the mouthparts. If the antenna is caught, then grooming proceeds normally. Between 5 and 9 days, the ipsilateral mesothoracic leg starts to be used to place the antenna. The success of this new behaviour gradually improves but, unlike *Blatta,* no transfer from ipsilateral to contralateral mesothoracic leg seems to occur. These results indicate some degree of plasticity in the nervous system of cockroaches. This may vary from species to species, and seems to be greater in *Periplaneta* than in *Gromphadorhina*. The emergence of a new pattern of grooming could be due to trial-and-error learning or possibly some developmental process. This modification of grooming could be the result of a new antennal catching movement by one of the remaining legs; equally it could be the emergence of a new method of standing which allows one of the legs to be free for catching the antenna. Some evidence for *Blatta* and *Periplaneta* suggests that the appearance of a new antennal catching movement takes longer if the animal is tested on a glass surface as opposed to a blotting-paper one, showing that devising a new way of standing may be at least part of the process. This may explain the poor adaptability of *Gromphadorhina,* which is a very heavy cockroach.

Time

To set up 10 animals and observe them daily will require 3 hours a day. Observation should be carried out for at least one week and preferably 2-2½ weeks. Additional time will be required for writing up.

8.3 Courtship of *Nauphoeta cinerea*

Purpose

This exercise allows the observation of a distinctive and relatively easily elicited courtship behaviour. It requires careful observation and description of the behaviour sequence by students, and introduces the problems of the stimulus-response sequence of the courtship which culminates in copulation and of the change in receptivity of the females.

Preparation

A sufficiently large colony must be maintained to supply the newly moulted cockroaches required for the exercise. Inspection of the colony once or twice a day for the week preceding the class will be needed to separate the newly moulted animals.

> 6-7 days before the class:
>   Collect enough newly moulted females to provide 2 per pair of students.
> 1-2 days before the class:
>   (a) Collect enough newly moulted females to provide 2 per pair of students.
>   (b) Separate from the colony enough mature females to provide 1 per pair of students.
> 1 day before the class:
>   Place the cockroaches into containers as described under *Materials.*

Materials

> per pair of students
>
> 1. 2 adult virgin cockroaches, 6-7 days old in a jar with lid.
> 2. 2 adult virgin cockroaches, 1-2 days old in a jar with a lid.
> 3. 1 mated female cockroach in a jar with a lid.
> 4. 5 adult male cockroaches.
> 5. 4 10-cm diameter crystallizing dishes.
> 6. Vaseline (petroleum jelly).

D

Methods

At the start of the class the crystallizing dishes should be liberally smeared with Vaseline around the inside rim to stop cockroaches escaping during observation. Each of the four dishes should then have one virgin female placed in it, and the animals should be left for about 5 minutes to settle down.

A male should then be introduced into one of the dishes containing a 6-7 day virgin female, and the behaviour of the pair observed. The test should be ended after 15 minutes, whether or not mating has been achieved. Similar observations should be made using the other 6-7 day virgin female, both the 1-2 day ones, and the mated female using a separate male for each. The behaviour of the animals may be recorded in note form and/or summarized on a score sheet such as Table 8.3.1.

With 6-7 day virgin females: On being placed in the dish the male will soon contact the female and in response to the female's presence, will raise his wings and tegmina (figure 8.3.1). He will then swing round

Figure 8.3.1 Male *Nauphoeta* shown with wings raised in courtship display.

to bring his posterior end towards the female. She will then begin to feed on a secretion from the male's tergum (figure 8.3.2). Gradually she moves forwards, mounting the male's back, still feeding until her genitalia are close to those of the male. Locking of the genitalia then occurs, and the female dismounts, the pair remaining *in copula* facing in opposite directions (Roth and Dateo; 1966). This courtship behaviour is very similar to that shown by the German cockroach *Blatella germanica,* and wing raising by males during

Figure 8.3.2 Female *Nauphoeta* shown feeding from tergal secretion of male. Male immobile in wings raised display.

courtship is also shown by *Periplaneta americana* (Cornwell, 1968).

With 1-2 day virgin females: Males approach these females and raise their wings; females usually respond by palpating the terga of the males, but not mounting far enough for mating to occur. After a number of unsuccessful mating attempts, the male will probably lower the wings and begin to show a quite different kind of behaviour. He elongates his abdomen and

begins to vibrate or shudder, at the same time producing a faint but audible noise. This stridulation is achieved by rubbing a ridged area on the underside of the pronotum against the costal wing vein (Hartman and Roth, 1967) and it is given by males to unreceptive females; however, the function of the stridulation is still obscure since it has not been demonstrated to have any effect on females over the long or short term.

With the mated female: As with the 1-2 day females, the male will show the courtship behaviour of wing raising. The female may respond by feeding from the posterior end of his terga but courtship will not proceed further. In this situation also the male may show some stridulation.

To summarize the findings, it will be observed that females are at first unreceptive but by 6 days after imaginal moult have become receptive; on mating, however, receptivity is lost again. There must

Table 8.3.1 Results that may be expected from male and female *Nauphoeta* using females of two different ages. The results are shown on a score sheet of a kind that could be used in the class.

(*a*) Behaviour of males towards females of different ages during courtship.

	Age of females (days)			
	1-2		6-7	
Behaviour pattern	Female 1	Female 2	Female 1	Female 2
Wing raising Stridulation Copulation	✔ ✔ ✘	✔ ✔ ✘	✔ ✘ ✔	✔ ✘ ✔

(*b*) Behaviour of females of different ages towards males during courtship.

	Age of females (days)			
	1-2		6-7	
Behaviour pattern	Female 1	Female 2	Female 1	Female 2
Dorsal feeding Copulation	✔ ✘	✔ ✘	✔ ✔	✔ ✔

therefore be internal changes affecting female receptivity.

The onset of receptivity in *Nauphoeta* was studied by Roth and Barth (1964) who concluded that it was controlled by some neurosecretory mechanisms not involving the corpora allata. The loss of receptivity in mated females was found to be due either to the presence of a male spermatophore in the female reproductive tract or an ootheca in the uterus (Roth and Dateo, 1966).

When the female is receptive and courtship does occur, it can be seen to be a series of reciprocal responses between male and female, culminating in copulation. When the male locates the female, he responds by raising his wings and turning his abdomen towards her. His response does not seem to be to a specific female odour, since he will sometimes wing-raise in response to contact with another male. When the male raises his wings, the female responds to a secretion on the back of the male by sexual feeding. This eventually brings her into a position where the response of the male unites the pair in copulation.

Time

A demonstration of the courtship behaviour using a single male and receptive female would take 10 minutes. Class observations using five females per pair would take 1½-2 hours. Shortened versions of the exercise could, however, be carried out.

8.4. Aggressive behaviour of *Nauphoeta cinerea*

Purpose

This exercise allows the observation of cockroach threat and appeasement postures, as well as overt fighting. It introduces questions about the development of aggression and of territoriality. It requires careful observation and description of behaviour.

Preparation

From 3 days before the class: Inspect the colony once or twice a day to collect newly moulted adult males. Mark them with a paint spot and place each in a separate container. Collect enough to provide 1 per pair of students.

The day before the class:

(a) Set up one 'resident' cockroach in a 15-cm diameter crystallizing dish with a Vaseline-smeared rim to prevent escape. Mark the cockroach with an identifying paint spot, and provide food and water in the dish. Set up enough to provide one resident cockroach per student pair.
(b) Set up one 'intruder' cockroach in a 15-cm diameter crystallizing dish for each pair of students in the same manner as for the residents.

Materials

Per pair of students
1. One resident adult male cockroach in a 15-cm diameter crystallizing dish
2. One intruder adult male cockroach in 15-cm diameter crystallizing dish
3. One newly moulted adult male cockroach in a separate container
4. One large nymph in separate container

Methods

The intruder male cockroach should be taken taken from its jar and dropped into the dish containing the resident. The behaviour of the two animals should then be observed and recorded:

(a) in terms of the fighting patterns. (Do they bite, butt, grapple, etc.?)
(b) in terms of the postures adopted by winning and losing animals.

This may be recorded descriptively or by filling in a score sheet such as that shown in Table 8.4.1. After one arrival has clearly become the victor, the vanquished should be removed, the newly moulted male introduced into the dish in its place, and the behaviour of the two males observed and recorded.

It will be seen that, when the intruder male is placed with the resident, both animals adopt an *aggressive*

Table 8.4.1 The behaviour patterns that may be expected from two adult male *Nauphoeta* during a fight which culminates in defeat for one of them. Results are shown on a score sheet of a kind that could be used in the class.

	Class of male	
Behaviour pattern	Winning	Losing
Courtship	✓	✓
Antennae fencing	✓	✓
Abdomen raising	✓	✓
Attack	✓	✓
Submission	✗	✓

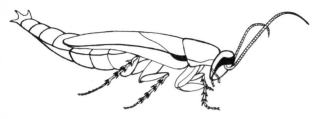

Figure 8.4.1 The aggressive posture shown by a male *Nauphoeta* towards a rival male (after Ewing, 1967).

posture (figure 8.4.1). When they approach one another, they will probably show rapid mutual *antennal fencing,* which is probably an aggressive behaviour, and is certainly more vigorous than normal antennal investigation. If the two animals are fairly well matched, fencing will probably be followed by fighting, during which the cockroaches will *butt* each other by charging with pronotum lowered, trying to toss the opponent onto its back. Sometimes males may *grapple,* their legs locked together, rolling over and over. While grappling, they also *bite* one another on the legs. The fight ends abruptly with one animal adopting the subordinate *posture* (figure 8.4.2), with

Figure 8.4.2 The subordinate posture shown by a male *Nauphoeta* towards a dominant rival (after Ewing, 1967).

body flattened, head lowered and antennae pressed onto the ground (Ewing, 1967). It may be the resident or intruder which wins the fight, but by pooling the class results it will be found that residents win significantly more frequently.

In the conflict between the resident and the newly moulted male, the latter will be found to show little or no aggressive behaviour and submits readily to the threats of the former. Similarly, the nymph shows no inclination to fight, indicating that factors determining aggression in males are not present in nymphs, and take three or more days after moulting before they influence the behaviour of adult males.

This exercise can be extended to observations on a colony situation, either as a demonstration or as a project study. Ewing (1972) found a dominance system in *Nauphoeta* colonies, with most dominant animals defending territories and collecting harems of females. A similar situation seems to occur in *Gromphadorhina brunneri* (Ziegler, 1972).

Time

A demonstration of a fight between resident and intruder males will take 10 minutes. Class observations using the four animals provided will take 20-40 minutes.

REFERENCES

Cornwell, P. B. (1968), *The Cockroach: A Laboratory Insect and an Industrial Pest,* Rentokil Library, Hutchinson.
Ewing, L. S. (1967), 'Fighting and Death from Stress in a Cockroach', *Science,* N.Y., 155: 1035-1036.
Ewing, L. S. (1972), 'Hierarchy and its relation to Territory in the Cockroach *Nauphoeta*', *Behaviour,* 42: 152-174.
Hartman, H. B. and Roth, L. M. (1967), 'Stridulation by the Cockroach *Nauphoeta cinerea* during Courtship Behaviour', *J. Insect. Physiol.,* 13: 579-586.
Luco, J. V. and Aranda, L. C. (1964), 'An Electrical Correlate to the Process of Learning', *Acta physiol. latinoam.,* 14: 274-288.
Roth, L. M. and Barth, R. H. (1964), 'The Control of Sexual Receptivity in Female Cockroaches', *J. Insect. Physiol.,* 10:965-975.
Roth, L. M. and Dateo, G. P. (1966), 'Sex Phenomena produced by Males of the Cockroach *Nauphoeta cinerea*', *J. Insect. Physiol.* 12: 255-265.
Ziegler, von R. (1972), 'Sexual and Territorialverhalten der Schabe *Gromphadorhina brunneri* Butler', *Z. Tierpsychol.,* 31: 531-541.

9. Locusts

INTRODUCTION

The most studied locust species are the desert locust *Schistocerca gregaria* and the migratory locust, *Locusta migratoria*. The contents of this Section refer more or less equally to both species, and in addition indicate the kind of behaviour which could be investigated in other locust (or even grasshopper) species, though not necessarily with the same result.

Both *S. gregaria* and *L. migratoria* are widely distributed in the Old World tropics. They are hemimetabolous insects, which means that their young (nymphal) stages resemble the adults except for the absence of wings. Uvarov (1921) was the first to discover that in *Locusta* two quite distinct forms of the species occurred at the nymphal stage: a solitary green form, and a black-and-yellow gregarious form. The appearance and also behaviour of a nymph depends on the proximity of other nymphs in the preceding instars. Solitary nymphs become solitary non-swarming adults, whereas gregarious nymphs become swarming adults covering large distances and causing considerable crop damage. This same pattern of development first described for *Locusta* is now known to be true for *Schistocerca*.

Locusts may be maintained in laboratory colony without too much difficulty, provided that quite high room temperatures can be maintained (28-30°C). No special humidity arrangements are required. Metal box cages with a Perspex front and a bulb to supplement the room heating are commercially available. Nymphs and adults should be fed on fresh green vegetation, such as lettuce and wheat germ; they should also have beakers of clean sand about 10 cm deep in which to lay their eggs. Locusts are also available from biological suppliers. All individuals either raised in your own colony or supplied will, of course, be of the gregarious form. The method of rearing solitary forms is described in 9.5. For all exercises except 9.4 it is slightly better to use *Schistocerca* rather than *Locusta,* because the former are more placid in the laboratory situation; *Locusta* are, however, preferable for 9.4 to contrast the behaviour of solitary and gregarious forms.

EXPERIMENTAL WORK

9.1. Feeding and food specificity

Purpose
Probably the most well-known feature of locusts is their apparently voracious and catholic appetite. This exercise gives an opportunity of observing locust feeding, of considering the stimuli which provoke feeding, and the location of the sense organs involved.

Preparation
The locusts should be deprived of food for 6-12 hours prior to the experiment.

The room should be made as warm as possible for the start of the exercise.

Materials

Per pair of students:
1. Two adult locusts
2. Two pieces of cotton thread about 40 cm long
3. One bench lamp

Per class:
1. Lettuce leaves
2. Fresh grass
3. Wheat germ (Bemax).

Methods

Firstly a slip knot should be tied in the cotton thread and the loop slipped over the head of the locust and tightened over the top of the pronotum and *behind* the prothoracic legs (figure 9.1.1). The locust may then be allowed to move freely over the bench, restrained only by its 'lead'. It should be kept near to the bench lamp to keep it warm;

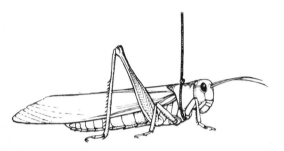

Figure 9.1.1 Loop of thread tightened round the pronotum and behind the prothoracic legs to secure the locust

A blade of grass should then be placed in front of (but not in contact with) the locust, and the behaviour of the animal observed. The sequence of events will be that the locust lowers its antennae to touch the grass lightly. The animal then holds the grass blade between the tarsi of the prothoracic legs and lowers its head on to the blade. The maxillary and labial palps then rapidly and lightly drum over the surface of the blade before the animal starts chewing. Having observed this feeding for not more than a minute, the grass blade should be withdrawn, so as not to satiate the animal.

Students will probably immediately recognize that touching the grass with antennae indicates the presence of chemoreceptors on the antennae concerned with feeding. Not so obvious is the presence of chemoreceptors on the tarsi, which probably give information about the nature of the food while the locust holds the grass. Additional information is also provided by the maxillary and labial palps. (A good account of these and other sense organs of the mouth parts is given by Chapman, 1974.) Another point to mention which may not be obvious to students is that,

in order to approach the food source, the locust must have some way of detecting it, either by visual or olfactory cues. Hungry locusts will move up-wind in response to the smell of crushed grass, showing that airborne olfactory cues do play a role.

The locusts should then be offered other kinds of food, such as lettuce, cabbage and wheat germ (Bemax) and will be found to eat these also, indicating a fairly non-specific food preference. This type of feeding is in marked contrast with insects, such as some caterpillars and aphids, where there is a single species of host plant.

Time

20-30 minutes, without discussion time

9.2. Single compounds as feeding stimulants

Purpose

Since the diet of locusts is rather non-specific, it is of interest to know what single compounds, such as sugars and amino acids, will initiate feeding. This exercise describes how this may be tested.

Preparation

The locust should be deprived of food for 6 to 12 hours prior to the experiment.

The solutions described under *Materials* should be made before the class (but not more than 24 hours before) to prevent contamination by micro-organisms.

Materials

Per pair of students:

1. Two locusts
2. Two 1-lb Kilner jars, with net covering the top.
3. About 6 cm of elder pith.

Per class:

1. Glucose solution (125 milli-molar)
2. Fructose solution (125 milli-molar)
3. Sucrose solution (125 milli-molar)
4. Xylose solution (125 milli-molar)
5. L-argenine solution (125 milli-molar)
6. L-alanine solution (125 milli-molar)
7. L-methionine solution (125 milli-molar)
8. Distilled water (control solution)
9. Tissues.
(Other sugars and amino acids may be tried as available.)

Methods

As there are more solutions than locusts, each locust should be tested with only one material. The results may be compiled as a class result at the end of the exercise.

The pith should be cut into 1-cm lengths and 3 pieces soaked in one of the solutions. The pith should be squeezed to make sure it has taken up the solution; it should then be lightly dabbed with a tissue to remove excess moisture before being placed in a Kilner jar. One locust should then be placed in the jar and the lid fastened. The reason for having a single locust to a container is that they are inclined to eat each other, which rather interferes with the results. After 3-4 hours the amount of pith eaten by each locust should be recorded on a ten-point scale where 10 = 100% of pith eaten. The class results should then be compiled to give the readings for all test solutions.

Chapman (1974) gives the results for *Locusta* in a comparable experiment which indicates that glucose, fructose, maltose and sucrose are good feeding stimulants, L-methionine slightly so, and xylose and L-alanine not stimulating at all. This picture seems to be broadly true for *Schistocerca* as well.

This exercise has been described as a class exercise involving 20 or more students, but it could be done by one or two students as a project investigation. In that situation there would be opportunity not only for testing various substances but also their concentration, and observing the effect of (say) amputating parts of antennae or palps. Chapman (1974) shows a rapid increase in feeding rate induced by fructose above a concentration of 25 milli-molar, followed by a rapid decline above 625 – so interesting results might be expected here.

The main point to bring out in discussion concerning this exercise is that of interpreting such experiments with single substances. A negative result with one substance does not necessarily mean that it is unimportant in providing feeding stimuli, but rather that alone it is not capable of stimulating feeding; it is possible that it may have a marked effect in enhancing feeding in combination with another stimulant.

Time

As a class exercise this may be set up quite quickly, certainly within 30 minutes. It must then, however, be left for 3 or 4 hours, then a further 40 minutes is required to compile and record the results.

As a project for one or two pairs of students it could take up 1-2 hours a day for 2-10 weeks.

9.3. Methods of locomotion

Purpose

This is a simple exercise to demonstrate that the locust has three distinct methods of locomotion, and shows the appropriate morphological adaptations for such methods of locomotion.

Preparation

Make sure the laboratory is as warm as possible.

Materials

Per pair of students:
1. One adult locust
2. About 40 cm cotton thread
3. A small ball of cotton wool
4. One bench lamp
5. One Pasteur pipette with rubber bulb
6. One retort stand with clamp

Methods

The locust should have a cotton loop placed round its thorax as described in 9.1. Methods.

Walking: If the locust is allowed to wander over the bench (but near the bench lamp so that it keeps warm), the animal will show its characteristic pattern of walking. Students may be asked to look for the pattern of leg movements suggested for the stick insects in Section 10.2, so the methods will not be repeated here. The locust is, however, appreciably less satisfactory than the stick insect; (*a*) because the leg movements are more rapid and (*b*) because they are less regular. The prothoracic legs do alternate left and right, and they are 180° out of phase with the mesothoracic legs; however, the metathoracic legs, particularly during slow walking, may miss a whole cycle or move both together. This irregular pattern of metathoracic leg movement is due to their specialization as jumping legs.

Jumping: If a finger is pressed gently down on the posterior of its abdomen, the locust will usually jump. This can be seen to be due to a rapid extension of the metathoracic legs. Prior to the jump, the locust adopts a characteristic position with the long slender tibia tucked right up against the massive femur (figure 9.3.1). In this position the extensor muscles in the femur wind up the tension, which is released by a trigger mechanism to produce the jump. The jump of the locust is, therefore, not the result of sudden muscular contraction as it is in the human for example.

Figure 9.3.1 Locust in position adopted immediately before the jump with tibia tucked against the large muscular femur.

Flight: If a locust is lifted off the ground by its thread, the wings will begin to beat in flight. If the locust is suspended from the arm of a retort clamp and allowed to continue to fly, it will stop immediately if a small ball of cotton wool is placed in its feet. When the ball is removed again, flight recommences. This allows students to deduce that there must be touch receptors on the tarsi, and that stimulation of these touch receptors inhibits flight.

The locust will not, however, fly indefinitely in this position and, after a minute or less, stops. A new burst of flying can be initiated by blowing on the animal from the anterior. By waiting till the locust stops flying and then directing puffs of air onto different parts of the body with a Pasteur pipette, it is possible to demonstrate that puffs of air directed onto the head between the eyes most readily elicit a burst of flight. The initiation of flight by wind is due to the stimulation of patches of wind-sensitive hairs on the head, first described by Weis-Fogh (1949) (figure 9.3.2). Having observed this, students may be asked to predict what will be the result of blowing on the head of the locust when the cotton wool ball is being held. They may think of a 'competition' hypothesis where strong stimulation of wind receptors overrides inhibi-

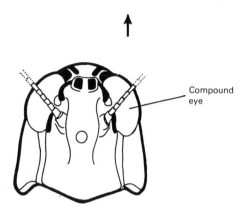

Figure 9.3.2 Head of locust seen from above. Black areas indicate the positions of patches of wind-sensitive hairs on the front of the head and along the upper edges of the compound eyes (after Weis-Fogh, 1949)

tion from the tarsi, or an 'exclusion' hypothesis where wind stimulation is unable to initiate flight so long as the tarsi are stimulated. This simple experiment demonstrates that the latter is true.

Time
All aspects of locomotion described here could be demonstrated by a teacher in about 20 minutes and could be carried out by a class in 40-50 minutes.

9.4. The mechanism of flight

Purpose
The behaviour of flight is the repetition of relatively simple movements, but the fact that this enables the animal to fly is in itself remarkable, and the frequent repetition of the movement allows investigation of the behaviour.

Preparation
None

Materials

Per pair of students:

1. Three adult locusts
2. Two 5th instar nymphs
3. Two 3rd or 4th instar nymphs
4. A small fan heater on an adjustable stand
5. A strobe light with flash rate from 1 to about 250 per second

6. A retort stand and clamp (latter with tilt adjustment)
7. 10 cm of copper tube 5 mm external diameter
8. A spirit lamp
9. A thermometer
10. 10 g soft wax
11. Metal spatula

Per class:
Adhesive tape

Methods

Flight of adult: A thread is fastened round the thorax of the locust as described in 9.1. The slip knot should be positioned exactly dorsally. The thread should then be passed through the copper tube and fastened to the top of the tube with adhesive tape, so that the pronotum of the locust is held gently against the bottom of the tube. A small piece of soft wax is then softened in the fingers and placed in a ring round the base of the tube. The locust is then welded to the tube by melting the wax ring with a spatula heated in the flame of the spirit lamp. The low melting-point of the wax and the good heat-conducting properties of the copper tube ensure that the locust does not get burned.

The fan heater should be turned on and a spot found in the airstream where the temperature is about 30°C. The locust should then be clamped in that position facing into the airstream (figure 9.4.1). The strobe light should be placed immediately beneath the flying locust and the flash interval adjusted so as to freeze the wing beat cycle (about 15/second). The reading of the flash rate on the strobe scale gives the wing beat frequency.

By *slowing down* the flash rate slightly, the greatly slowed wing beat cycle may be observed. It is then possible to observe that

(a) Forewings lead the hind wings by about $\frac{1}{4}$ of a cycle (Pringle, 1975).
(b) The forewing always has a slightly positive angle of attack relative to the airflow, and the angle is larger on the recovery stroke, thereby reducing the air resistance.
(c) There is a characteristic leg posture during flight (figure 9.4.1) with the prothoracic held flexed just beside the head; the mesothoracic and metathoracic legs are extended backwards along the body. (The positions of the meso and metathoracic legs are rather unreliable in this situation, but the prothoracic leg position should be readily seen.)

Tilting: If the animal is tilted relative to the airflow, it can be seen that the pronation of the forewing relative

Figure 9.4.1 Mock-up of a locust suspended in a warm airstream. This situation causes a locust to show sustained flight with legs held as shown in a characteristic flight position. Note the cotton thread passed round the prothorax, and the pronotum stuck to the metal tube with wax.

to the body is increased as the head is gradually raised. The form of the hind wing beat is substantially unaffected by variation in body angle. The overall wingbeat frequency will be seen to remain the same; the animal therefore maintains lift and attempts to correct the tilt only by varying the angle of attack of the wings.

Rotating in horizontal plane: An animal held in a horizontal position may be rotated in that plane so that the airstream no longer strikes it directly head on. Weis-Fogh (1949) found that animals in this situation tried to head into the wind again and that this was due to asymmetric stimulation of the wind-sensitive hairs on the head. At the level of analysis of this exercise, all that can really be observed is that the wingbeat frequency remains almost unchanged and the two sides remain in phase. These observations together suggest that differences in wing angle and wingbeat path correct yaw in flight, indeed close examination of the wingbeat suggests some asymmetry of this kind.

Removing part of the wings: Having observed the

flight of the intact locust, about half the wing length should be removed on both sides and the animal reflown. It will be observed that the wingbeat frequency is only slightly altered. This shows that loss of wing area, which has altered the physical properties of the wingbeat system, has not, however, changed the rate of muscular contraction. This suggests that the centrally generated commands determine the contraction frequency of the wings, and that these commands are uninfluenced by any sensory feedback (Wilson, 1968 *a*). This type of mechanism for determining wingbeat frequency is called *synchronous* and is found not only in the Orthoptera, but also in the Odonata and Lepidoptera (Wilson, 1968 *b*). In the *asynchronous* type of wingbeat, found in Diptera, Hymenoptera and Coleoptera, the beat frequency is a property of the muscles themselves, so that shortening the wings may result in as much as three-fold increase in wingbeat frequency.

'Flying nymphs': If a 5th instar nymph of *Schistocerca* and *Locusta* is attached to a rod, and held facing into the warm airstream, it may show the full-flight position and quite reliably shows the prothoracic-leg flight position. This indicates that part of the reflex response of flight has developed fully when the wings are still buds. Third and fourth instar nymphs do not show this response. This 'flight posture' of nymphs has been described by Bently and Hoy (1970) for crickets, and by Kutsch (1974) for nymphs of *Schistocerca* from 2nd instar onwards.

Time

This exercise is best suited to a practical class of 2-3 hours, because animals take time to set up, and some individuals turn out to be poor fliers. The results of this practical should optimally be conducted in conjunction with a discussion, because the results will not be very clear-cut, but chiefly because the exercise raises problems about the principles of flight in general, and the control and ontogeny of the flight mechanism in the locust in particular.

9.5. Social behaviour of solitary and gregarious nymphs

Purpose

It was pointed out in the introduction that in both *Locusta* and *Schistocerca* the appearance of the nymphs depends on whether they grow up in groups or isolated. Grouped rearing produces *gregarious* nymphs which are conspicuously coloured yellow and black and have a saddle-shaped pronotum (figure 9.5.1*b*). Rearing in isolation results in *solitary* nymphs, which are grey-green in colour and have a markedly ridged pronotum (figure 9.5.1*a*). These differences in appearance are accompanied by differences in behaviour which may be investigated or demonstrated in a class situation.

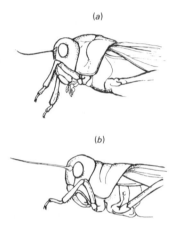

Figure 9.5.1 Side view of anterior of *Locusta* to show difference in appearance of solitary and gregarious forms. (*a*) Solitary form showing dorsal surface of the pronotum convex and ridged. (*b*) Gregarious form showing dorsal surface of the pronotum concave and saddle shaped (after Imms, 1957).

Preparation

Normal colony rearing of locusts produces only nymphs of the gregarious form, and it is as yet not possible to obtain solitary nymphs from suppliers, in Britain at least; therefore, if it is planned to use solitary nymphs for a class exercise they must be specially reared. This is, unfortunately, a long and laborious business, but worth it because, when students are able to see the solitary and gregarious nymphs together, they are not surprised that they were originally classified as different species.

6-8 weeks before the practical class

The rearing of solitary nymphs of sufficient size (4th and 5th instar) will take 6-8 weeks.

Newly emerged first instar nymphs should be placed singly in jars with a net covering over the top, and green vegetation such as lettuce provided. Isolation of the nymphs is necessary, since tactile and olfactory communication between nymphs induces gregarious appearance and behaviour. The effects of this air-borne pheromone should be further reduced by spacing the jars 25 cm or more apart and keeping them in a well-ventilated room. It is not necessary to isolate the locusts visually.

The confinement of a single locust in a jar could itself result in a sufficient accumulation of pheromone within the jar to induce gregarious characters; in order to avoid this, each nymph should be removed from its jar every day and placed temporarily in a specimen tube. Its jar must be completely cleaned of the previous day's food plus faeces. The inside of the jar should then be thoroughly wiped with a paper tissue. The locust must be replaced in its jar with fresh food, and the specimen tube cleaned thoroughly for use the following day. Separate specimen tubes should be used for each nymph. All nymphs should be fed and cleaned out every day, including weekends, for at least the first two instars; this not only prevents the accumulation of pheromone but also cuts down mortality of the rather delicate young nymphs due to starvation or desiccation. Even so mortality will be rather high (30-40%).

48 *hours before the practical class*

All nymphs to be used in the class, whether solitary or gregarious, should be deprived of food.

30-60 *minutes before the practical class*

The solitary nymphs should be placed in with the gregarious ones.

Materials
For class exercise of orientation and approach.

Per pair of students:
1. Ten gregarious nymphs
2. One or two solitary nymphs
3. One plastic washing-up basin 30 cm across with a transparent Perspex or acetate lid
4. A bench lamp with 100-watt bulb
5. A thermometer

For class demonstration of marching:

1. A locust cage
2. 50-60 gregarious nymphs
3. 2-3 solitary nymphs
4. Some lettuce leaves or other locust food

Methods

Class exercise: Orientation and approach by gregarious and solitary nymphs

Each pair of students should be provided with a plastic basin covered with a transparent lid and containing 10 gregarious and one or two solitary nymphs. The bench lamp should be placed close over the basin to keep the temperature high. Once the nymphs have settled down, they will begin to walk about, and may be observed and their behaviour recorded. It is, however, necessary to keep fairly quiet and not to bang the surface of the bench, as this causes the locusts either to freeze or jump up and hit themselves against the lid.

Gregarious nymphs will be seen to show bursts of fast walking, followed by immobility; sometimes they will be seen walking parallel or line ahead. Sometimes one gregarious locust will be seen to start walking apparently in response to another locust walking past, suggesting a facilitating effect. Gregarious nymphs when motionless will often position themselves with their bodies parallel to one another.

Solitary nymphs may also walk about quite actively, but their walking is never obviously facilitated by that of other nymphs. They do not co-ordinate their walking with that of other nymphs, and do not show obvious orientation to other locusts while walking or at rest.

There is another aspect of the behaviour of solitary and gregarious nymphs in this situation on which we have little information, but which might be worth following up. If a gregarious nymph approaches another gregarious nymph, the former usually stops when its antennae touch the latter. If the approach is head to head, the two individuals show a bout of antennal fencing with their heads a few millimetres apart.

If a gregarious nymph approaches a solitary one, the response of the latter to antennal contact is to back away slightly or even walk away. When a solitary nymph approaches a gregarious one, however, it may walk right up to the other until its head actually touches before stopping. These observations seem to suggest that gregarious nymphs respond socially to one another in an organized way which is not fully developed in solitary nymphs.

This is an interesting exercise for students at undergraduate level, but might not prove suitable at lower levels because the behaviour differences between the two types of nymphs are not very clear cut, and because the locusts are disturbed by bangs on the bench and the like.

Demonstration: Marching by gregarious and solitary nymphs

The ten hungry gregarious nymphs provided per pair of students in the class exercise above seem to be insufficient in number to induce continuous marching. Since this is rather a dramatic and convincing behaviour, it is worth setting up as a class demonstration. 50 or 60 gregarious nymphs should be placed in a standard locust cage of 40 cm square floor dimensions, with light bulb inside to give a temperature of 28-30°C, and Perspex front for easy viewing. All locusts should be 48 hours starved, and the solitary nymphs should have been introduced into the cage about 30 minutes before the practical.

By the beginning of the class, obvious marching should be occurring. This is a quite characteristic kind of behaviour. The locusts walk fast and persistently, their bodies held somewhat above the ground and their heads well up. In addition to the fast walking, they give little hops of 3-5 cm, sometimes in a run of 2 or 3 like a bouncing ball. Although some locusts will be standing still, the majority will be circulating rapidly round the floor of the cage, all in the same direction. To show that the smallness of the hops is not due to exhaustion, but is a specific marching behaviour, it is only necessary to give the cage a sharp tap; this causes many locusts to show escape jumps of 15 cm or more.

It will be noticed that only the gregarious nymphs show the marching behaviour and that, although solitary nymphs may walk about, they do not follow the circulating column of gregarious animals, nor do they show the characteristic raised posture of marching gregarious nymphs.

To demonstrate that the marching behaviour is induced by hunger, a few lettuce leaves may be dropped into the cage. These will be voraciously eaten in about 20 minutes, after which the locusts will become quite immobile.

Since the marching behaviour is quite impressive to students, it is worth while demonstrating with gregarious nymphs alone, without going to the bother of raising solitary nymphs.

Time

The class exercise will take 30-50 minutes, depending on the detail required.

The marching demonstration is best shown to a few students at a time, taking 5-10 minutes each time. After this the locusts should be allowed 20 minutes feeding time, and should then be inspected, again to confirm that all marching has stopped.

REFERENCES

Bently, D. R. and Hoy, R. R. (1970), 'Postembryonic Development of Adult Motor Patterns in Crickets', *Science*, N. Y. 170: 1409-1411.

Chapman, R. F. (1974), *Feeding in Leaf Eating Insects*, Oxford Biology Readers, No. 69, Oxford University Press, London.

Kutsch, W. (1974), 'The Development of the Flight Pattern in Locusts', in *Experimental Analysis of Insect Behaviour*, Ed. L. Barton Browne, Springer-Verlag, Berlin; New York.

Pringle, J. W. S. (1975), *Insect Flight*, Oxford Biology Readers, No. 52, Oxford University Press, London.

Uvarov, B. P. (1921), 'A Revision of the Genus *Locusta*, L. (= Pachytylus, Fiels.) with a new Theory as to the Periodicity and Migrations of Locusts', *Bull. ent. Res.*, 12: 135-163.

Weis-Fogh, T. (1949), 'An Aerodynamic Sense Organ Stimulating and Regulating Flight in Locusts', *Nature*, London, 164: 873-874.

Wilson, D. M. (1968 *a*), 'The Flight Control System of the Locust', *Sci. Am.*, 218 (5): 83-90.

Wilson, D. M. (1968 *b*), 'The Nervous Control of Insect Flight and Related Behaviour', *Advances in Insect Physiology* 5, Academic Press, London.

10. Stick Insects

INTRODUCTION

The Phasmids, stick and leaf insects, have a predominantly tropical distribution. They are all herbivorous and often have very remarkable anatomical modications to enable them to resemble the twigs or leaves among which they dwell. They are largely nocturnal. This Section refers to one species of stick insect *Carausius morosus* because it is commonly kept in laboratory culture or classrooms throughout Britain. Apart from its elongated body, its appearance is not very exotic, but the exercises described for it should work for many other stick insect species if they are available. *Carausius morosus* has wingless adults. It reproduces parthenogenetically, as do a number of stick insect species. Laboratory colonies are therefore almost exclusively of females, males occurring only rarely. It occurs in two fairly distinct colour forms, green and brown. Brown individuals are capable of some colour change towards green, depending on a variety of environmental variables, e.g. light, humidity and temperature. Similar colour changes are also shown by other species.

Carausius is easily maintained in the laboratory or classroom on a diet of fresh ivy leaves *Hedera helix* or privet leaves *Ligustum vulgare,* both of which are available through the year. A large plastic box with some gauze over the top is an adequate container, and no special heating arrangements are required, a room temperature of between 19°C and 23°C being optimal. There are usually six nymphal instars, but growth is rather slow, from egg to adult taking 3-4 months.

EXPERIMENTAL WORK

10.1. Dissection of nervous system

Purpose
To show the segmental arrangement of ganglia with paired connections between them in the thorax and abdomen region. This is an important exercise prior to the behavioural ones described later, where students are required to think how co-ordination of one leg with another is achieved.

Preparation
None

Materials

Per student or pair:
1. 1 final instar stick insect
2. 1 dissecting microscope
3. 1 pair of fine forceps
4. 1 pair of fine dissection scissors
5. 50 g Plasticine
6. 8-10 pins
7. 1 Pasteur pipette
8. 70% alcohol

Methods
This is a very easy nervous system dissection, appreciably more easy than the cockroach or locust.

(a) Take a more-or-less full-grown animal and kill it in a killing bottle.

(b) Make a flat Plasticine platform, put it under the binocular microscope, and fasten the stick insect dorsal side up to the platform by its legs, using blobs of Plasticine.

(c) Insert the scissors in the mid-dorsal line, behind the 2nd or 3rd abdominal segment, and cut forward to the head. This can be done very quickly.

(d) Lift out the gut.

(e) Flood the body cavity with 70% alcohol to make the nervous system turn from the transparent to opaque white.

(f) Pull away the few fragments of connective tissue and fat which remain in the cavity, and the segmental ganglia and paired connectives are clearly revealed (figure 10.1.1)

(The dissection to expose the head ganglia is difficult because of densely packed muscles in the head.)

Time
20-40 minutes

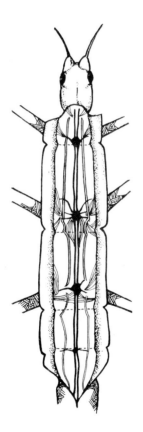

Figure 10.1.1 Dorsal view inside the thoracic region of the stick insect body cavity after removal of the gut. The positions of the thoracic ganglia can be seen corresponding to the positions of the legs and joined by paired connectives.

10.2. Leg co-ordination during walking

Purpose
To illustrate that the sequence of leg movements during walking is not haphazard but patterned. This patterning of behaviour must be an expression of the underlying organization of the nervous system, and for the legs to be co-ordinated with one another there must be nervous communication across a single ganglion and between ganglia (see Exercise 10.1.) Compare with Exercise 2.1 on the dissection of the nervous system of the earthworm and the experimental investigation of its locomotion.

Preparation
None

Materials

Per student:
1. One, second or third instar stick insect
2. One fairly fine paint brush

Methods
Persuade a stick insect to walk, and observe pattern of leg movement. This is most regular in earlier instars, so they are probably better to use.

The first thing to be observed is that when one leg is lifted up and moves forward, the leg of the same segment on the opposite side remains on the ground holding the animal up. This suggests some reciprocal inhibition across the ganglion to prevent both legs extending at the same time; this can be contrasted with jumping in grasshoppers, where both metathoracic legs extend in synchrony.

The second feature to be observed is that three legs are brought forward simultaneously (front left, middle right, back left) while the body is supported by the triangle of the other three legs. The three raised legs are then lowered in their new position, and the other legs raised and brought forward. The animal therefore progresses in a gait of so-called 'alternating triangles'.

Having observed this regularity of walking pattern, the problem may then be raised with students as to whether the walking pattern is an entirely centrally generated pattern, or whether it is substantially feedback controlled. Of course, not much can be done in an introductory laboratory class to clarify the issue one way or the other, but there is one brief exercise that can easily be done to illustrate the effect of sensory feedback on stepping pattern. This is most effectively done on a larger animal whose stepping rate is a bit slower than for young animals. The procedure is to allow the insect to walk, and then touch the tarsus of the *mesothoracic* leg just as the metathoracic leg on the same side is being brought forward; this requires a bit of skilful timing, but what will be observed is that the metathoracic leg touches the ground and immediately makes a little step backward. This illustrates a number of simple but important points. The location of sense organs on the tarsus, the presence of sensory nerves from tarsi to ganglion, the presence of interneurones between meso and metathoracic ganglia and of motor neurones from metathoracic ganglion. It also demonstrates that the stepping pattern may be altered by sensory input and is therefore not exclusively centrally patterned. A more detailed review on insect locomotion, including the stick insect may be found in Rockstein (1974).

Time
10-25 minutes

10.3. Leg raising patterns in standing insect

Purpose
One of the problems for students of observing the pattern of leg movements during walking is that the stepping cycle occurs so rapidly it is impossible for a record to be made of the sequence of events. An exercise which illustrates the same points of central patterning, communication between ganglia and influence of sensory input, is that of leg raising in the standing insect. It has the additional advantages over the walking exercise that it allows a flexible, more experimental approach by students, and provides time for scoring of the behaviour response.

Preparation
None

Materials

Per student:
1. One final instar stick insect
2. One fairly fine paint brush
3. Data scoring sheets (if required)

Methods
Allow a stick insect to walk on the bench until it comes to a stop. If it is reluctant to stop, then persuade it to crawl up a vertical surface; this usually stops it. The tarsus of one leg is then touched lightly with the paint brush, which causes that leg to be raised and held in the air. If another foot is touched, that leg will be raised. Clearly this process cannot continue indefinitely without some legs being lowered again! The student is therefore required to study which combinations of raised legs are easy, and which are difficult or impossible to obtain. It will be found, for example, that both front legs raised together, a common resting position in *Carausius,* is quite easy to obtain. The triangles of raised legs observed during walking may also be obtained, but with less regularity.

At all school levels it is probably advisable to provide prepared data sheets to score which leg was stimulated, and which legs were raised or lowered as a result of the stimulus application. At the undergraduate level, students will, after ten minutes' initial observation, devise their own scoring system — a useful exercise in the problem of how to record behavioural data.

Time
2-40 minutes

10.4 Cutting circumoesophageal connectives

Purpose
Sections 10.2 and 10.3 suggest to the student that there is some kind of central mechanism which regulates stepping pattern or leg-raising pattern. This exercise investigates the problem of whether different ganglia contain, as it were, different parts of the programme. This is done by providing animals with their supra-oesophageal ganglia effectively removed, so that their walking and leg-raising behaviour may be compared with intact animals.

Preparation
The operation, which is of cutting both circumoesophageal connectives, is best done the day before or at least three or four hours before the class begins. This is quite tricky to do, and therefore needs to be performed by the teacher. Since CO_2

anaesthetic produces rather erratic behaviour in stick insects, it is best to operate without anaesthetic. Hold the head of the stick insect between thumb and finger and, with a fine scalpel blade or razor blade splinter, cut forward along the line shown in figure 10.4.1, so as to cut through one of the connectives. Turn the animal round and cut the connective on the other side. The successful cutting of the second connective is often accompanied by an upward curling of the abdomen, followed by a convulsive twitch.

Figure 10.4.1 Side view of the head of a stick insect, showing the position of the head ganglia and connectives. The dashed line indicates the position of the cut to pass through one of the circumoesophageal connectives.

Materials

Per student:
1. One intact final instar stick insect
2. One final instar insect prepared (operated) as above
3. One fairly fine paint brush

Methods

Students are given an operated animal and asked to observe its locomotion and leg raising in comparison with an intact animal. The following differences should be observed. Operated animals:

    (a) walk slowly;
    (b) walk continuously or, if they stop, recommence walking very readily;
    (c) drag their bodies along the ground;
    (d) have a more or less normal stepping pattern;
    (e) raise their legs when tarsi are touched, but almost immediately lower them again.

These observations serve to demonstrate that the stepping pattern and foot raising are still present as in unoperated animals, showing that the supraoesophageal ganglion is not essential for their expression. The loss of the ganglia has, however, caused a general loss of muscular tone, as shown by the dragging body and drooping legs. The continued walking also suggests a disinhibition of locomotion resulting from removal of the ganglion.

Time
20-35 minutes

REFERENCE

Rockstein, M. (Ed.) (1974), 'The Physiology of Insecta', 2nd ed., Vol. 3, Ch. 5: Hughes, G. M. and Mill, P. J., *Locomotion: Terrestrial,* Academic Press, N.Y.

11. Caddis Larvae

INTRODUCTION

Trichoptera have a worldwide distribution. There are about 150 species of caddis larvae in Britain alone which build portable houses; however, we have only a rudimentary knowledge of their house building and that for only a handful of these species. Any work undertaken on house building is likely to produce at least some completely original results. The projects outlined here are, however, illustrated with results obtained from three species: *Lepidostoma hirtum, Silo pallipes* and *Athripsodes atterrimus* (figure 11.0.1, Hansell, 1973 and 1974), but it is hoped that the suggestions given may encourage people to try similar studies on the numerous other species available.

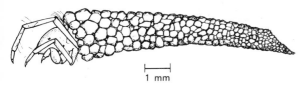

├─────────┤
1 mm

Figure 11.0.1 Third instar larva of *Athripsodes atterimus* showing house of sand grains (after Hansell, 1972)

The species being studied may be identified using the book *Caddis Larvae* by Hickin (Hutchinson, 1967).

Caddis larvae are widely distributed in freshwater habitats throughout the World, so a locality where they can be collected should be readily to hand for all except perhaps city dwellers. Larvae will not, however, be found in water which is obviously polluted or on a substrate which is mainly fine mud. Therefore, the places to look for larvae are in the reeds and water weed at the edges of unpolluted streams or ponds, or on gravel or stony substrates in streams, at the edges of lakes or in gravel pits. In places where they do occur, it is usually in quite reasonable numbers; sometimes they are very plentiful.

In streams they can sometimes be seen as little specks sitting on large boulders, and can either be picked off or swept off with the hand into a net held immediately downstream. On gravel substrates, a net with a durable bag and sturdy metal frame about one foot in diameter on a long broom-pole handle should be used to sweep up the surface layer of gravel on which the larvae are sitting; such a net is also ideal for sweeping through vegetation at the edges of streams or ponds. The way to pick out the camouflaged larvae from the gravel or vegetation is to spread the material thinly over the floor or a large shallow white-enamel dish full of water; wait for half a minute, then the caddis larvae will be seen as they lean out of their houses and start to walk.

Larvae may be kept for days, or even weeks, in aquarium tanks or plastic basins at room temperature. The tanks should be well aerated, although this is not necessary for crystallizing dishes containing individual larvae. Larvae should not be brought straight from cold outside to warm room temperature (20°C or more), but should be gradually adjusted to it over a few days; only then should any experiments be performed on them. The floor or the container should be covered with sand or gravel, and for those species such as *Silo* and *Agapetus* that feed on encrusting algae and diatoms, large stones taken from the habitat should be provided. For vegetation feeders, such as *Lepidostoma* or *Anabolia,* appropriate vegetation should be included. Carnivorous or scavenging species, such as *Phryganea* or *Athripsodes* survive quite well with no special attention.

EXPERIMENTAL WORK

11.1. Observation of the 'life cycle' of house building

Purpose
Most caddis larvae have five larval instars. Most of the species that build houses do so from the first instar, extending the house at the anterior as they grow and cutting off sections of the house at the posterior. Considering all the species of caddis larvae in Britain, the detailed house structure and building behaviour of the larvae throughout their lives are known, for, almost literally, half a dozen of them. The problem is therefore to study in detail the house building throughout larval life of a species as yet unstudied.

Preparation
None

Materials

Per pair of students:
1. A good supply of larvae of the species being studied.
2. A student monocular microscope with moving stage.
3. A stereomicroscope.
4. 20-30 crystallizing dishes, either 7 cm or 12 cm diameter.
5. 5-10 Petri dishes of about 8cm diameter.

Per class:
1. A light trap (ideal but not essential).
2. Long-handled collection nets for larvae.

Methods
Eggs are laid as a mass, embedded in a gelatinous envelope. They may be found attached to vegetation or stones, in or just beside ponds and streams. Many species of adult caddis can be caught in a light trap, and females of some of the species carry their egg masses around partially extruded. If these females are placed in a small cage with a Petri dish lined with wet filter paper, they will lay eggs on the filter paper:

the advantage of collecting eggs in this way is that the adult may be identified immediately using a key such as that in *The British Caddis Flies* (Mosely, 1939), whereas identification from larvae cannot be done with certainty until the fifth instar using a key such as that in *Caddis Larvae* (Hickin, 1967).

Eggs should be kept in small dishes of water. When the larvae hatch, they should be transferred to dishes containing the same sorts of potential building material as those that occur in the natural habitat: aquatic vegetation, dead leaves, sand, etc. Only one larva should be placed in each dish, so that the life history of individual larvae may then be studied. Twenty larvae should provide an adequate sample for study.

If eggs cannot be found, it is still very worth while to collect first and second-instar larvae from their natural habitat, to place them in individual dishes, and then commence observations; however, in this case a check should be made in the fifth instar to make sure that all individuals are of the same species.

The following measurements should be made on all larvae, twice a week:

> (*a*) house length
> (*b*) anterior outside house width
> (*c*) posterior outside house width
> (*d*) larval head width

When larvae are very small, the easiest method of performing these measurements is using a student microscope with a calibrated moving stage. If a hairline is put in the eyepiece, the dimensions of the house and larva to the nearest 0.1 mm may be obtained. When the larva reaches the fourth or fifth instar, measurements with a ruler may be possible.

Using these simple methods:

> (*a*) a moult may be identified by the sudden great increase in head width;
> (*b*) addition to the house may be identified by the increase in house length;
> (*c*) the cutting off of a posterior section of the house may be identified by a sudden substantial drop in house length, and an increase in posterior house width.

An illustration of the sort of interesting results that may come from such a project is given by a study on *Lepidostoma hirtum* (figure 11.1.1) (Hansell, 1972). This shows that during any instar the larva extended

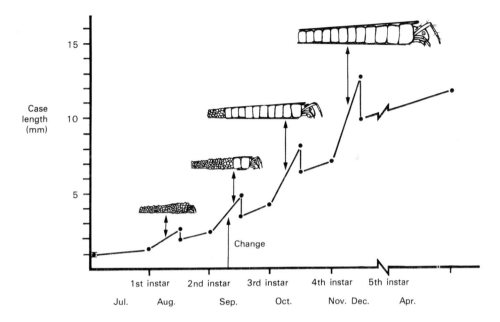

Figure 11.1.1 Change in house length during the larval life of *Lepidostoma hirtum*. Line drawings indicate the proportion of sand and leaf in the house at four positions during larval life. The point in the third instar of the change from attachment of sand grains to leaf panels is also shown (after Hansell, 1972).

its house continually in small amounts, but that the larva turned round and cut off a large posterior section of the house only once in each instar, this occurring about halfway between two moults. In the third instar, between the moult and the cut-off, the larva changed from constructing a sand-grain house of tubular section to constructing a house of rectangular pieces of leaf arranged with one row for each of four sides.

Time

This exercise can only be performed as an extended project.

As many species lay eggs in the late summer and autumn, study may start with the academic year. Larvae will probably pupate the following spring. This study would therefore require about two hours' work, twice a week for seven months.

11.2. Investigation into the detail of the house structure

Purpose

The house of a caddis larva is a partial record of that animal's behaviour. The behaviour of an animal is organized so that a sequence of behaviour patterns will produce a useful end result. A caddis house is the result of such a sequence of behaviour patterns and therefore contains enduring information on the organization of those patterns. The problem is to examine the house structure in detail to extract that information.

Note: This section, even in a shortened form is a very useful preliminary to any project which concentrates mainly on the experimental study of the organization of house building behaviour (e.g. 11.6-8) since useful clues can be obtained on the experimental methods which are likely to give interesting results.

Preparation

None

Materials

Per pair of students:
1. A good supply of larvae of the species being studied.
2. A student monocular microscope with moving stage.
3. A stereomicroscope.
4. Ten crystallizing dishes of 7 cm diameter or larger.

Per class:
*1. Microtome for cutting wax sections.
*2. The basic requirements for light microscope histology, i.e. facilities for staining and embedding.

* These techniques are only appropriate to vegetation houses and are desirable extra techniques rather than essential.

Methods

Larvae of any instar collected from the wild may be used. Fifth-instar larvae may be more convenient because of their larger size; as was shown for *Lepidostoma* in. 11.1, house architecture may change from one instar to another. Therefore species should be identified and, if possible, the instar. Actual observations will vary, depending on the type of house being studied, but these are a few suggestions:

(a) Are particles of different materials (e.g. some sand, some leaf)?

(b) If there are different materials, are they of few types (e.g. sand grains and dead leaves in *Lepidostoma*) or many types?

(c) In houses made of one type of material, is there more than one clearly recognizable particle size?

(d) Are particles differing either in size or type of material placed in different parts of the house (e.g. one at the anterior, the other at the posterior; one for the roof and floor, another for the sides)?

(e) Are particles regular in shape?

(f) If particles are regular, are they oriented in a particular way with respect to the house (e.g. largest dimension across the house)?

(g) Do houses of different larvae of the same species differ greatly or only a little from one another?

The techniques used in these investigations may be various. Observations on the number of particles per house might be done by eye. Dimensions of house particles can be measured with the moving stage of a microscope as in 11.1. Information on the arrangement of panels in leaf houses can be obtained by embedding whole houses in wax and cutting sections. A feature which will be observed for many species (e.g. *Athripsodes,* Hansell, 1973) is that a house is wider and incorporates larger particles at the anterior end. The reason for this is that the larva built the anterior part of the house several weeks after the posterior part, during which time it has grown. This conclusion is reinforced by the observation in 11.4 that reconstructed houses, which are built in a few days, show no taper.

Observations on the fifth-instar house of *Silo pallipes* (Hansell, 1968 *a* and *c*) showed that the roof

Figure 11.2.1 House of *Silo pallipes* seen from above: arrow indicates anterior. The roof can be seen to be composed of small particles, and each side of two large particles. Both small and large particles are larger at the house anterior (after Hansell, 1974).

and floor of the house were composed of small sand grains. On each side of the house were two much larger stone particles. Both large and small particles were smaller at the posterior than the anterior of the house (figure 11.2.1). The large particles were oriented with their longest dimension along the axis of the house, and their smallest across it. Small particles, however, were oriented with their longest axis across the axis of the house and their smallest vertically through it. Both were therefore oriented to cover the maximum area, but large particles were oriented to give the house length rather than height. The large particles also leaned inwards, so that the roof was narrower than the floor (figure 11.2.2).

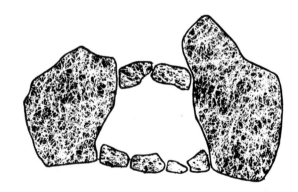

Figure 11.2.2 Transverse section through the house of *Silo pallipes* showing that roof and floor particles are laid flat so as to cover the maximum area and that the large side particles lean inwards.

Time

This exercise is most suitable as an extended project; however, the time spent on it may be very variable.

Thirty minutes' observation may be sufficient to get over the idea that caddis larvae select materials rather than just gather them indiscriminately, and that selection criteria may change. There is, however, enough information in most species to provide interesting study for 2-4 hours a week for two months.

11.3 Observation on the repair of damaged houses

Purpose

If a small part of the anterior of a house is removed, the response of the larva is to replace the damaged part. Since the particles replaced are similar to those removed, it may be assumed that the behaviour of the larva is similar to that shown during normal house extension. This is therefore a method of inducing and studying the behaviour of attaching new particles to the front of the house. The problem is to observe the behaviour shown by a larva during particle replacement, describe the different stages in particle selection, and determine how these stages result in the selection of a suitable particle.

Preparation

The dishes should be set up with building materials and water at least two hours before the experiment, and then stirred just before the experiment to remove the tiny bubbles which have formed in the dish. This is important, because the bubbles tend to become attached to the hairs of larvae and cause them distress.

Larvae which have part of the house removed may spend an hour or more chewing away the loose silk in the damaged region; therefore, if class time is limited, larvae should be prepared for the class about two hours before, and kept without building material until the start of observations.

Materials

Per pair of students:
1. A good supply of larvae of the species being studied.
2. A stereomicroscope or magnifying lens depending on size of the species and level of study.
3. Five or more 7-cm-diameter crystallizing dishes.
4. A pair of fine forceps.

Methods

About one quarter of the house should be removed from the anterior using fine forceps, under magnification if necessary. The larva should then be placed in a dish containing building material of the type removed from the house. The larva should then be observed in detail. Observations on not less than 10 animals will give a good idea of the behaviour and the similarity between animals.

The actual information recorded will vary, depending on whether the larva is selecting unmodifiable building units from the environment (sand grain builders) or cutting building units from the environment (vegetation builders); however, here are some suggestions:

Sand grain builders

(*a*) How does the larva locate particles?
(*b*) How does the larva attach particles to the house?
(*c*) Are there recognizable stages in locating and attaching particles to the house, e.g. the larva in different positions or using different legs at different stages?
(*d*) Are some particles rejected before attachment to the house?
(*e*) How do rejected particles differ from accepted ones?

Vegetation builders:

(*a*) How does the larva locate particles?
(*b*) How does the larva attach particles to the house?
(*c*) Is the particle fashioned in stages? How do the particles differ in the different stages?

It is possible to remove particles of sand and vegetation from larvae while they are actually manipulating them, using fine forceps. Larvae are briefly disturbed by this, but resume building behaviour. Particles collected during the observation period can then be measured later.

The removal of a small portion of the roof of *Silo pallipes* causes it to lean out of its house scratching particles towards it with rapid movements of the pro- and meso-thoracic legs (figure 11.3.1). Only some particles contacted in this way are picked up: those which are, undergo manipulation with all legs, and some of these are rejected (figure 11.3.2). Any remaining particle is held up to the damaged area and turned round and over once more. Some particles are rejected at this stage; some are fitted (figure 11.3.3).

Measurement of particles rejected at the different

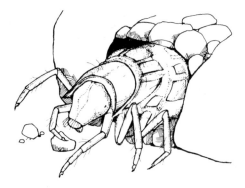

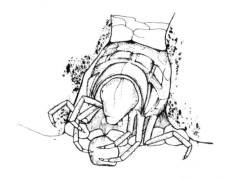

Figure 11.3.1 Scratch test: *Silo* larva leaning out of its house making scratching movements with the pro- and mesothoracic legs.

Figure 11.3.2 Manipulation test: *Silo* larva picking up a particle for manipulation with all legs.

stages showed that scratching operated as a test, where all particles above a threshold size were picked up, and all those below it were ignored (figure 11.3.4). As a result of the manipulation of particles, larvae dropped all particles above a certain size, giving a narrower range of sizes which were then tried in various positions on the anterior house rim; some of these were then fitted, others rejected (Hansell, 1968 *b*).

Lepidostoma hirtum repairs its house with cut-out leaf panels. These are obtained by the larva holding onto the edge of a leaf and cutting directly inwards with its mandibles. The cut extends in a straight line for a while and then stops. The larva then starts a second cut at the the edge of the leaf; this extends into the leaf, and then turns through 90° to meet the first

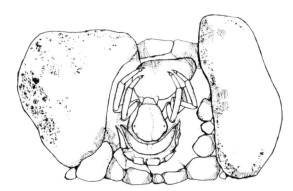

Figure 11.3.3 Experimental fit test: *Silo* larva testing the particles in various orientations at the position of roof damage.

Figure 11.3.4 The percentage of sand grains accepted by *Silo* larvae at each of the three test stages plotted against the largest dimensions of the sand grains in mm.

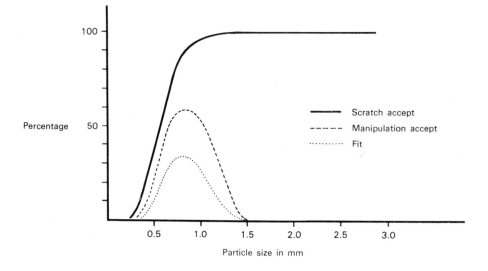

cut. The larva then removes the rectangular panel and attaches it to the anterior of the house.

Time

One hour of observation using three or four animals is probably the minimum to give an idea that there is a more or less stereotyped pattern in particle selection. For an extended project, however, it is probably necessary to observe ten animals for one hour each, as a preliminary to an experimental study. For a more detailed study on the size and shape of particles accepted at different stages, about 20 animals should be studied in detail.

11.4 Observations on house reconstruction

Purpose

Most of the caddis species so far studied seem to build their first house soon after hatching; the larva then extends the house as it grows. If a larva is pushed out of its house, it finds itself naked for the first time since soon after it hatched. This situation usually results in the construction of a house either slightly or very different from the house from which the larva was removed. This demonstrates that different stimulus situations may induce the larva to show different patterns of behaviour. The problem is: what is the response of your species to this situation, and how can it be explained?

Preparation

The dishes should be set up with building materials at least 2 hours before the experiment and then stirred just before the experiment to remove the tiny air bubbles (see Preparation 11.3).

Silo, *Athripsodes* and *Lepidostoma* all take about an hour from being pushed out of their houses to the establishment of the initial collar of the house; after this building is rapid. If class time is limited, larvae should be set up either right at the start of the class or before. Where only the completely reconstructed house is being studied, 24 hours should be allowed for reconstruction.

Materials

Per pair of students:
1. A good supply of larvae of the species being studied.

2. A stereomicroscope or magnifying lens, depending on the size of the species and level of study.
3. Five or more 7-cm-diameter crystallizing dishes.

Per class:
1. A box of pins.

Methods

There are two ways of carrying out this study. The first is to take 10-20 larvae; push them out of their houses by inserting a blunt pin in the posterior end of the house; place each larva in a dish containing a variety of building materials, including both sand and vegetation; and leave for 24 hours. The larvae are then removed and their reconstructed houses compared in detail with those from which they were removed. The second is to observe the actual reconstruction behaviour from the time of removal of the larva from the house. This can then be compared with the building behaviour shown by a larva with a portion of the anterior of the house removed, as in 11.3, where house repair is apparently the same as normal house extension. When an *Athripsodes* larva is pushed out of its house and allowed to reconstruct a house on a sand substrate, the larva starts by silking a lump of sand grains together; it then pulls a string of sand grains over its abdomen to form a collar round the body. Sand grains are then added to the anterior of the collar to form the new house tube, which, unlike the original house, is untapered. The size of the sand grains is less regular than that of the small particles of the original house, suggesting a relaxation of the criteria for the acceptance of particles. However, the anterior of the reconstructed house is more regular than the posterior suggesting that, as the body of the larva becomes more covered, so the criteria for particle acceptance become narrower.

When a fifth-instar larva of *Silo pallipes* is removed from its house, it reconstructs a house which is a simple tube of more-or-less uniformly-sized sand grains. There are no large particles on the sides, indicating that in this situation, only one of its two particle-selection processes operates.

Time

It is a worthwhile exercise to observe house reconstruction or study reconstructed houses for an hour only; however, a systematic comparison between, say, 20 original and 20 reconstructed houses and particle dimensions would take 2-3 hours' work a week for 4-5 weeks.

11.5 Experimental investigation of entry into empty houses

Purpose
If larvae are pushed out of their houses and provided with empty ones lying on the substrate, they will enter them very readily. This raises the problems:

(a) How does a larva recognize a house from the outside?
(b) Can a larva recognize the anterior from the posterior end of the house?
(c) Can a larva distinguish a suitable from an unsuitable house?

Preparation
The floors of two crystallizing dishes should be thickly covered with molten wax. When this has hardened, the dish should be filled with water at least 2 hours before the practical. Just before the practical, the inside of the dish should be wiped with a paint brush to remove air bubbles.

Materials

Per pair of students
1. Ten caddis larvae.
2. One dissecting microscope or magnifying lens, depending on the size of the species and level of study.
3. Two 7-cm-diameter crystallizing dishes, the floors of which are covered with a layer of wax.
4. A fine paint brush for removing bubbles from dishes.
5. A stop watch or clock.

Per class:
1. A box of pins.

Methods
If three larvae are pushed out of their houses and placed in the wax-floored dish, together with about eight empty houses of varying size, they respond immediately they contact an empty house; so observation may start immediately.

A naked *Athripsodes* larva, presented with an empty house, grabs hold of it and walks along it to the opening at either the wide or narrow end, and then attempts to enter. If it enters the anterior (wide) end, it will then turn round inside the house to face forwards before walking off with the house. If it enters the posterior (narrow) end of the house, it crawls through to the anterior and walks off without turning round. A larva entering a house which is either too big or too small continues to examine empty houses when they are contacted, eventually leaving the first to enter another. Details of the response of a number of species of larvae to this situation are given by Merrill (1969).

These observations demonstrate that:

(a) a larva can recognize complete houses from the outside;
(b) a larva can distinguish the anterior from the posterior of the house, at least after entry;
(c) a larva can distinguish a suitable from an unsuitable house.

This represents the control situation in which such things as the following may be recorded:

(a) the time taken to enter a house;
(b) the end through which the larva enters;
(c) the manner of entry (head first; tail first);
(d) the behaviour of the larva after entry.

After this, houses may then be experimentally modified and the behaviour observed again. The sorts of modification that may be tried are:

(a) providing houses which are either too short or too long;
(b) providing houses which are too wide or too narrow;
(c) providing houses which are blocked at one or other, or both ends.

Time
The entry into empty houses and turning round inside the house can be demonstrated in ten minutes. In 2-3 hours, 10-15 larvae can be systematically observed in more than one experimental situation. An interesting experimental project could be sustained for 8 or 9 weeks working 2-3 hours a week.

11.6. Experimental modification of house structure

Purpose
The houses of some species have special architectural features. For example, in *Lepidostoma* the roof at the anterior projects in front of the sides, and those in front of the floor. In *Silo,* large particles are attached to the side of the house, and small particles to the roof

and floor. The problem is whether a larva remains aware of these architectural features after they have been incorporated into the house, and whether it can restore them if they are selectively removed.

Preparation
The dishes should be set up with building materials at least 2 hours before the experiment, and then stirred just before the experiment to remove the tiny air bubbles (see Preparation 11.3).

Materials

> Per pair of students:
> 1. A good supply of larvae of the species being studied.
> 2. One stereomicroscope or magnifying lens, depending on the size of the species and the level of study.
> 3. About 20 crystallizing dishes, 7 cm or 12 cm diameter.

Methods
This depends very much on the house of the species involved, but here are a few suggestions:

> (a) If the anterior rim of the house is such that part of it projects more than another, cut off part of the anterior of the house so that all sides are level. Then see if the larva repairs the house so as to restore the unevenness of the anterior rim.
> (b) If the house is composed of particles of more than one kind, take away parts of one kind or the other, e.g. take away an anterior large particle from the house of *Silo pallipes* to see if it is replaced by a large particle. Another example is the house of *Anabolia nervosa* (figure 11.6.1) which consists of a tube of small sand grains and pieces of vegetation, along the side of which are laid 2 to 4 twigs, which often project beyond both ends of the house. One or all of these twigs can be removed, or some or all of the anterior or posterior projections.

Figure 11.6.1 House of *Anabolia nervosa* showing tubular structure of sand grains and pieces of vegetation, with huge twigs attached which protect beyond the ends of the house.

Experiments should be conducted by modifying the house, placing the larva in a dish containing building materials, and allowing at least 24 hours for reconstruction. The particles replaced should then be compared in detail with those removed. About 20 larvae should be used for each experiment to get a clear picture of the effect.

When small particles are removed from the anterior of the roof of *Silo*, they are replaced by similar *small* particles. This demonstrates an ability to detect the change in the house and respond appropriately. If a *large* particle is removed from the anterior of a *Silo* house, it is also replaced by *small* particles. This demonstrates an ability to detect but not respond in the same way to the change.

No one has looked systematically at the consequences of experimental modification of the twigs on the house of *Anabolia nervosa*, although we have observed casually that if all the twigs are removed, at least some of them are replaced. This certainly looks like an interesting problem for someone to have a look at.

Time
This is only suitable for an extended project approach. Each experiment will take 2-3 hours to prepare, using 20 animals per experiment. The results of each experiment may take a further 3 weeks at 2 hours a week to work out.

11.7. Experimental alteration of building materials

Purpose
This section is most suited to those larvae which build with sand grains, because they have to select particles ready made from the environment. They must therefore sort through many items in the environment in order to find a few correct ones. The problem is, how do they recognize a particle as 'correct'?

Preparation
Dishes should be set up with building materials at least 2 hours before the experiment, and then stirred just before the experiment to remove the tiny air bubbles (see Preparation 11.3).

Special building materials, such as lead particles, may take several hours to prepare, and so attention should be paid to this problem well before the exercise is planned to start.

E

Materials

Per pair of students:
1. A good supply of larvae of the species being studied.
2. A stereomicroscope or magnifying lens, depending on the size of the species and the level of study.
3. About 20 crystallizing dishes, 7 cm or 12 cm diameter.

Methods

A small part of the anterior of the house of a larva should be removed, and the larva placed in a dish with its normal building material; this is a control larva. Another larva has the same amount of house removed, but is given building material which differs in a specific way from that given to the control. Larvae should be left for at least 24 hours, and a detailed comparison made between particles fitted by experimentals and controls. The object of the experiments may differ from species to species, but here are two suggestions:

(a) To test if larvae are selecting for weight of particles, or size, or both. Control larvae are given a range of sizes of sand grains; experimental larvae a similar range of sizes of lead particles, prepared by chopping up a piece of lead with a knife.

(b) If larvae normally fit flat particles, it is possible to determine if particles must have a particular ratio of length to width in order to be accepted. This can be done by offering larvae a range of sizes of glass spheres (commercially available) at diameters of 4.0 mm down to about 0.5 mm.

There is some confusion in the literature as to whether larvae select sand grains for weight or size, but *Silo pallipes* chooses the same-sized particles whether they are lead or sand, demonstrating that it selects only for size. It would be interesting to know how many other species also select in this way.

Given glass spheres, *Silo* will select ones whose diameter is intermediate between the largest and smallest diameter of the flat particles it normally accepts. This seems to demonstrate that in the absence of particles of the right *shape*, it selects particles of the same sort of *volume* to normal particles.

Time

As in 11.6, this is an exercise most suited to an extended project and so would yield best results if given 2 or 3 hours a week for 3-4 weeks.

11.8 Experimental modification of the larva

Purpose

If part of the house of a larva is removed experimentally, and the larva subsequently replaces it, this demonstrates:

(a) The larva has sense organs which detect removal of parts of the house.
(b) The larva has sense organs which tell it that it has completed the repair.
(c) The larva has sense organs which, during the repair, tell it that it is fitting the right particle in the right place.

The problem is to locate these sense organs and determine what information they are providing. One way of doing this is to remove parts of the larva where the sense organs might be, and observe how the building behaviour is altered.

Preparation

Dishes should be set up with building materials at least 2 hours before the experiment, and then stirred just before the experiment to remove the tiny air bubbles (see Preparation 11.3).

Materials

Per pair of students:
1. A good supply of larvae of the species being studied.
2. A stereomicroscope.
3. About 20 crystallizing dishes, 7 cm or 12 cm diameter.

Per class:
1. One or more pairs of fine scissors (required for leg amputation). Iridectomy scissors are ideal but expensive.
2. One or more pairs of watchmaker's forceps.
3. Menthol crystals.

Methods

The possible function of the eyes may be investigated simply by removing a portion of the anterior of the house of a larva of the control group, putting the larva in a dish with building materials, and allowing it to repair its house in continuous light, but allowing a larva of the experimental group, similarly prepared, to repair its house in continuous darkness. At least 24 hours should be allowed for house repair, and up to 20 larvae should be tested in each group.

To test the possible function of the legs in the measurement of particle size or shape, a larva of the experimental group should be anaesthetized in saturated menthol solution and one pair of its legs amputated. A larva of the control group should be anaesthetized but not operated upon. A portion of the anterior of the house should be removed from the houses of both groups, and a comparison made between the particles replaced by the two groups. Three experimental groups may be run:

(*a*) Prothoracic pair of legs removed.
(*b*) Mesothoracic pair of legs removed.
(*c*) Metathoracic pair of legs removed.

Up to 20 larvae of the experimental group should be compared with the same number of controls; however, the same controls may be common to more than one experimental group.

To test the possible function of hairs as touch receptors, groups of them may be removed from anaesthetized larvae by plucking them off at the base with watchmaker's forceps. House repairs of 20 larvae with hairs removed may then be compared with those of 20 controls.

The function of the anal hooks may be investigated by snipping them off at the base with microscissors and comparing the house repair of 20 such larvae with that of 20 controls. Again both experimental and control larvae should be anaesthetized.

The result of allowing larvae to rebuild in the dark will most likely be to show that houses of experimentals and controls are the same, thus demonstrating that eyes are not important.

A similar result will probably be found for the removal of any one pair of legs, but in this case, it either means that the pair of legs is not providing useful sensory information or that, in the absence of that pair of legs, the remaining legs can provide the necessary information.

Very little work has been done on removal of hairs, and any result here is of interest; however, it is true of *Lepidostoma* that removal of the anal hooks and associated hairs results in the replacement of more house than was removed (Hansell, 1972). This shows that these structures give information as to whether the posterior end of the larva is covered and that, in the absence of information from these sense organs, house building continues too long.

Time
As in 11.6 and 11.7 this is really only suitable for an extended project study. To set up 20 leg or anal hook amputees plus controls takes 4-5 hours. The results of such an experiment will then take almost 20 further hours to analyse.

11.9. Investigation of the predators of caddis larvae

Purpose
It seems reasonable to assume that caddis larvae build houses to fulfil some function. One possible function is as a protection against predators. If this is true, the house could act as a protection in one or both of two ways: concealment and mechanical protection. The problem is therefore whether the house of the species we are studying protects the larva against predators and, if so, how?

Preparation
None

Materials

Per student or pair:
1. A good supply of larvae of the species being studied.
2. At least two aquarium tanks, 30 cm x 25 cm x 25 cm, or larger.
3. Ten or more crystallizing dishes, 12 cm diameter or larger.
4. A stereomicroscope.

Per class or group:
1. One or more long-handled collecting net.
2. Fish gill nets.
3. A rowing boat.

Methods
First it is necessary to identify possible predators. This means making collections with long-handled nets and possibly also, fish-gill nets in the habitat where the larvae are found. The lengths or weight of the fish should be measured, the contents of the stomach examined, and the presence or absence of caddis larvae recorded. Invertebrate predators, such as dragonfly nymphs, feed by sucking in half-digested body fluids, so no information can be obtained from their guts; in this case evidence of predation must come from observation of the response of the potential predator to caddis larvae in an aquarium tank.

Identification of the predators is, however, only half the problem. The other half is how, if at all, do caddis houses protect the larvae against predators? Possible experiments to study this are:

(a) Provide different-sized fish with uniformly-sized caddis larvae. Observe which sizes of fish are successful and which are not. Observe and record the behaviour of both fish and caddis larvae in successful and unsuccessful predation attempts. Compare the sizes of fish taking larvae with the size of fish predating similar caddis larvae in the natural habitat.

(b) If certain caddis larvae are known to be eaten by particular fish, they can be presented to the fish in a tank on a concealing and a non-concealing background, and the time taken for the fish to locate the houses recorded.

There are a number of accounts in the literature of caddis larvae being found in the stomachs of fish; however, there are none that we know of, where an investigation has been made as to whether the house protects the larva from being eaten by certain kinds or sizes of fish or indeed any other predator. The general lack of knowledge therefore makes this study particularly interesting.

Time

This exercise can only be performed as an extended project. It will probably require collecting trips once every 2 or 3 weeks. The duration of these will, of course, vary depending on the proximity of a good habitat. Experiments should be interesting (because they deal with active animals) but time-consuming, because the results may not be clear cut. Probably 2-3 hours a week for 10-12 weeks would be necessary to get a consistent picture.

REFERENCES

Hansell, M. H. (1968a), 'The House Building Behaviour of the Caddis Fly Larva *Silo pallipes* Fabricius. I. The Structure of the House and Method of House Extension', *Anim. Behav.*, 16: 558-561.

Hansell, M. H. (1968b), 'The House Building Behaviour of the Caddis Fly Larva *Silo pallipes* Fabricius. II. Description and Analysis of the Selection of Small Particles', *ibid.*, 562-577.

Hansell, M. H. (1968c), 'The House Building Behaviour of the Caddis Fly Larva *Silo pallipes* Fabricius. III. The Selection of Large Particles', *ibid.*, 578-584.

Hansell, M. H. (1972), 'Case Building Behaviour of the Caddis Larva, *Lepidostoma hirtum*', *J. Zool.*, London, 167: 179-192.

Hansell, M. H. (1973), 'A Laboratory Practical on the House Building Behaviour of the Caddis Larvae', *Journal of Biological Education*, 7(3): 3-7.

Hansell, M. H. (1974), 'The House Building of Caddis Larvae: A Course of Projects for Schools', *ibid.*, 8(2): 88-98.

Hickin, N. E. (1967), *Caddis Larvae*, London, Hutchinson.

Merrill, D. (1969), 'The Stimulus for Case Building Activity in a Caddis Worm (Trichoptera)', *J. exp. Zool.*, 158: 123-130.

Mosley, M. E. (1939), *The British Caddis Flies*, London, Routledge.

12. Flour Beetles

INTRODUCTION

Flour beetles of the genus *Tribolium* (Coleoptera) are pests of stored grain and grain derivatives. They have been reported in flour, oatmeal, rice flour, patent breakfast cereals, chocolate, spices, certain nuts, and even occasionally as predacious on specimens in insect collections.

The two species commonly used in the laboratory, *Tribolium confusum* and *T. castaneum,* are particularly convenient experimental animals; they are slow-moving, easily handled, simple to culture; they cannot climb up vertical glass or Perspex walls, and rarely fly at normal laboratory temperatures.

Tribolium are best cultured in small milk bottles plugged with cotton wool, or covered plastic sandwich boxes. The culture medium can vary according to requirements, but a complete diet is provided by 12 parts whole wheat flour to 1 part yeast (by weight). The cultures should be kept in a constant-temperature room or incubator at 28-30°C, and should be cultured at 3-monthly intervals.

The eggs are about 0.4 mm long and 0.6 mm wide; the burrowing larvae pass through from 6-11 instars, depending on temperature and culture medium, and finally the pupae hatch into the small polished brown imagines which are 3.3-3.5 mm in length. At 28°C it takes approximately 30 days from egg to adult. The adults are best handled with a pooter or fine paint brush and strips of filter paper.

EXPERIMENTAL WORK

12.1 Responses of *Tribolium* to angle of substrate

Purpose

If a piece of paper (filter paper) is crumpled and placed on the top of the medium in a culture box, the flour beetles tend to aggregate on the highest ridges, folds and points. Students will quickly note this marked distribution and can be set the task of establishing the reason for it. Three variables are usually suggested: it could be a movement towards light, it could be a gravity response, or it might be a contact response. Since the lower and side edges of the paper can be seen to be comparatively free of beetles, the idea of a contact response or 'edge' response can be rejected. However, *Tribolium* do respond negatively to gravity and positively to light. This experiment is therefore suggested as a way of establishing that they do in fact move up an incline and, further, of investigating their sensitivity to slopes in a quantitative manner. Their response to light is investigated in 12.2.

Preparation

If the initial observations are to be made as suggested above, the crumpled paper should be introduced to the culture boxes an hour prior to the class. To accentuate the aggregations for introductory dicusssion we have occasionally put lamps over the boxes. This warms the animals, makes them more active, and they move up and towards the light more markedly.

Materials

> Per group of students:
> 1. Culture box of *Tribolium* as in Preparation above.
> 2. Bench lamp.
>
> Per pair of students:
> 1. Pooter
> 2. Fine paint brush
> 3. 2 sheets of filter paper
> 4. 2 glass or Perspex tanks, 50 cm × 20 cm × 10 cm
> 5. 4 sheets of sugar paper, 50 cm × 20 cm
> 6. 2 large protractors
> 7. Plumbline
> 8. Clamps and stand
> 9. 2 opaque black cloths for covering tanks
> 10. Small beaker for holding adult *Tribolium*
> 11. Spirit level.

Methods

Sheets of sugar paper, which must be absolutely flat, are fitted flush on the bottoms of the two tanks. The tanks must be rigid, and the bottoms perfectly flat. (Tanks can be made up with Perspex or, preferably, sheets of glass sealed with adhesive and edged with tape). One tank can be used level and the second can be at an angle of 45°. Use a spirit level for the first tank, and plumbline and protractor when clamping the second in position. This procedure is essential if any quantitative work is to follow the initial demonstration that the beetles do in fact tend to move up an inclined surface. Graham and Waterhouse (1964) have shown that *Tribolium* respond to angles of incline as small as 1°. They further demonstrated the need for a surface on which the beetles can move with ease. On a smooth Perspex surface the experimental animals tend to slip and aggregate, over a period of time, in the lowest part of the apparatus; hence the paper floor to the arena.

With care in setting up the tanks, students should be able to release 20 animals in the centre of the paper and demonstrate the sensitivity of *Tribolium* to very small angles of incline in darkness. It is suggested that this is dealt with before an investigation into their response to light, so that the experimenter is aware of at least one source of 'apparatus effects'.

Time

2 hours, or overnight

12.2. Light responses of *Tribolium*

Purpose

It is recommended that this exercise be preceded by an investigation into responses to surface angles as above (12.1). However, it can be carried out in its own right, perhaps as a part of a programme of investigations into light responses of a variety of species, as long as the tutor is aware of the possibility of very slight slopes in apparatus causing aggregations. This exercise is designed to demonstrate the photopositive response of *Tribolium* imagines.

Preparation
None

Materials

> Per student or pair of students:
> 1. Small beaker with 20 *Tribolium* adults
> 2. Glass or Perspex trough, 50 cm × 20 cm × 10 cm
> 3. 4 sheets of card or sugar paper, 50 cm × 20 cm
> 4. Pooter
> 5. Paint brush (fine)
> 6. Spirit level
> 7. Bench lamp or ray box
> 8. Opaque black cloth or black paper

Methods

Provided that the arena is set up so that it is perfectly level, the procedure is very simple. The sheet of card or paper is set flush on the bottom of the tank and checked to see that it is perfectly flat. Twenty *Tribolium* are released at the centre of the tank, and their distribution noted at intervals of three or four minutes. The tank is kept in very dim, diffuse light or darkness by shading it with cloth or paper between observations. Finally, half the tank can be uncovered, so that a bench lamp is illuminating one end, or a ray box can be shone through the end wall of the tank. There is normally a marked aggregation at the lighted end of the tank.

Time
1 hour

12.3. Humidity responses of *Tribolium* adults

Purpose
Although humidity is apparently not critical to the

survival of *Tribolium* in culture (they thrive anywhere between 25% and 80% R.H.) they do in fact respond to gradients of humidity. The present exercise investigates this response.

Preparation

The humidity choice chambers need to be set up for at least an hour before testing the animals, so that a humidity gradient is established. It is suggested that pretreated animals are used in at least part of the exercise. This means that a number of animals should be kept over KOH crystals in a desiccator for three days prior to testing. Likewise a batch can be kept over water for a similar period if required.

Materials

Per pair of students:
1. Humidity choice chamber
2. Container with 10 *Tribolium* adults
3. Pooter
4. Sheets of acetate film or polythene
5. Wax pencil
6. Outline plans of chamber
7. Stop watch or clock
8. Opaque cloth cover for chamber
9. Oil paints for marking animals
10. 3 mounted needles for putting paint spots on beetles
11. Spirit level.

Methods

The humidity chamber is essentially that used, for example, with Woodlice (Section 5.1). Special care should be taken to ensure that the arena material (fine net) is tightly stretched and perfectly flat and level. Check it with a spirit level (see 12.1).

The animals are introduced (say 10 individuals per test) and their distribution noted at five-minute intervals. Between observations the chamber must be shielded from light by a cloth or black-paper cover. The positions of the animals can quite conveniently be marked with a wax pencil on a circle of acetate or polythene sheet laid on the top of the chamber. Alternatively, numbers in the humid and dry sides can simply be recorded on a prepared worksheet. If the movements of individuals are to be followed, then a marking scheme must be devised. Oil paints do not interfere with the behaviour of the beetles, and can be used for colour coding the animals. Mounted needles are probably best for putting the minute spots of colour on the elytra. The results with beetles direct

from the culture bottle or sandwich box are variable, though in general more will tend to be found in the dry area than the humid one. When individuals are followed, the students will note that they spend a proportion of their time in the humid and the dry regions.

With pretreated specimens (over KOH for 3 days) there is an immediate imbalance in the distribution. Whereas normally the distribution is biased to the dry area, with the desiccated animals there is a noticeably larger proportion of animals in the humid area.

Time

1-3 hours or overnight

12.4. Food preferences in *Tribolium* species

Purpose

This is a very simple introductory exercise to introduce the students to the different species of *Tribolium* in a simple ecological context. These beetles are pests of stored cereals and similar substances, and this exercise is to study the choice or preference for different food substrates of *T. confusum* and *T. castaneum,* thereby leading to a partial understanding of their differing distributions in stored grain derivatives. This exercise can be used before 12.1 as a general introduction.

Preparation

Foods to be tested should be stored for at least 24 hours in the same humidity (e.g. in a desiccator)

Materials

Per student:
1. 1 container with 25 *T. castaneum* adults
2. 1 container with 25 *T. confusum* adults
3. Pooter or paint brush and filter paper
4. 3 or 4 large crystallizing dishes
5. 1 container 'Wheatfeed'
6. 1 container crushed oats
7. Cloths or paper to cover dishes
8. 1 tea strainer or sieve

Methods

The procedure is extremely simple. The Wheatfeed is passed through the sieve (tea strainer) and spread over half the floor of a crystallizing dish. The crushed oats, which have been stored with the Wheatfeed so that there is no humidity difference between them, are passed through the same sieve and spread over the other half of the floor of the dish. Twenty-five *T. confusum* are released in the dish and observations are made hourly (or daily if required). A second dish is set up as above and an equal number of *T. castaneum* introduced. Over a period of a day or two, students normally get a 2:1 preference for Wheatfeed with *T. castaneum,* while *T. confusum* shows no distinct preference.

There is obviously scope for further development of this method on open-ended lines. Apart from a range of common stored cereals being checked, an investigation into particle size is possible.

Time
Observations over 3 days or more

REFERENCES

Graham, W. M. and Waterhouse, F. L. (1964), 'The Distribution and Orientation of *Tribolium* on Inclines, and the Concept of Controls in Gradient Experiments', *Anim. Behav.,* 12: 368-373.
Park, T. (1934), 'Observations on the General Biology of the Flour Beetle, *Tribolium confusum*', *Rev. Biol.,* 9: 36-54.

13. Water Beetles

INTRODUCTION

The larger carnivorous water beetles and their larvae are superb subjects for the behaviour laboratory. They are attractive, readily obtainable, simple to house and maintain, and easily handled. Their movements and appearance, coupled with their ferocious carnivorous habits, make them immediately interesting to the younger as well as the more mature student. This Section suggests one or two simple introductory studies which may be of value to tutors trying to capture interest early in a course.

Probably the two most useful species are *Dytiscus marginalis,* the Great Diving Beetle and its smaller relative *Acilius sulcatus.* These form the basis of the work outlined here. But other dytiscids can be used and, outside the British Isles, equivalent species may well repay study and interesting comparative investigations result.

There are approximately 100 species of Dytiscidae in the British Isles, and nearly two and a half thousand species have been described from all over the world; but the group is usually considered Palaearctic in character.

As with all dytiscids, *D. marginalis* is exclusively carnivorous, both as larva and adult. Its eggs are laid in incisions made in submerged water plants. The larvae appear in the spring and, when fully grown, pupate in small oval cells in damp soil at the edge of the pond. The imagines emerge after about 20 days, but usually remain in the cells until their body surfaces are fully hardened, which may take a further 10 days. The sexes can be distinguished by the texture of the elytra; those of the male are smooth and glossy, while those of the female are finely lined or furrowed along their length. *A. sulcatus* has a similar life history; its eggs, however, are scattered singly on to the mud on the bottom of the pond. The larva of *Acilius* is easily recognized by its peculiar 'neck'. The first segment of the thorax is exceptionally long, and gives the larva this characteristic appearance (figure 13.0.1).

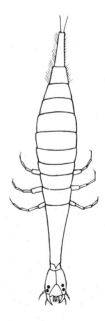

Figure 13.0.1 *Acilius* larva.

Dytiscid larvae and adults can be collected by net and scoop from ponds and canals. In winter it is often worth while scooping through mud from the pond bottom. *D. marginalis* adults frequently burrow into mud at this time of year. Most biological supply houses can provide larvae and adults of certain species, but may require prior notice.

In the laboratory, whether adults or larvae, individuals are probably best housed separately in small jars of pond water containing a little sand, pebbles and pieces of a pond plant such as *Elodea canadensis*. Adult beetles, however, can be kept together if desired, and larger aquaria should be used to study, for example, their breeding behaviour.

Since larvae and imagines require live food, adequate provision of suitable food species is important. Earthworms, *Enchytraeus, Tubifex,* tadpoles (when available) can form the basis of the diet of both young stages and the adults. Other pond animals, even young fish, will be readily accepted.

EXPERIMENTAL WORK

13.1. Simple observations on locomotion in *Dytiscus marginalis*

Purpose
This is a simple introductory exercise to familiarize the student with the animal and, in particular, the structure and functions of the limbs.

Preparation
None

Materials

Per student:
1. 1 preserved *Dytiscus* imago
2. 1 live *Dytiscus* imago
3. 1 preserved *Dytiscus* larva
4. 1 live *Dytiscus* larva
5. 2 small observation cells or beakers of pond water
6. 1 set dissecting instruments
7. 1 hand lens or low-power stereomicroscope
8. 1 bench lamp
9. 1 white tile

Per group or class:
One pair of mating adult beetles (if available).

Methods
The exercise is primarily one of observation and familiarization.

Larvae: Teaching patterns and methods vary, but we would suggest that initially the live larva be observed, then the preserved larva be dissected and examined and, finally, when the student is familiar with the structure of the larval appendages, the range of larval movements be analysed in the living animal again.

The student usually identifies six characteristic patterns of movement in the live larva.

1. The resting position on the floor of the container, or on weed, or when feeding.
2. Crawling.
3. Swimming, which in character differs little from crawling, except that the hair-fringed legs are obviously efficient oars.
4. The behaviour at the surface film as the larva takes in air. The last two abdominal segments and the small pair of terminal lobes are fringed hairs which enable the larva to hang head downwards from the surface film and take in air through the caudal spiracles.
5. 'Aggressive' stance. When alarmed, or mildly irritated by jolting the observation cell, or approached by (for example) a blunt seeker, the larva will take up a rather scorpion-like stance with its tail held upright.
6. When disturbed with a blunt seeker, the animal may show serpentine curving of the body, which causes it to move jerkily and speedily away from the instrument.

Imagines: Similarly, we suggest that initial observation of the movements of the imago should be followed by an examination of the preserved specimen, so that the student can return to the live animal with a detailed picture of the structure of the limbs in particular, a general appreciation of the streamlined form of the beetle and, perhaps, an appreciation of the fact that the imago is comparatively little modified for its life in water. (If desired, its general structure could be usefully compared with that of a terrestial beetle.) It should become clear with testing the animal out of water that it can indeed move and live on dry land and, although few of the animals fly under laboratory conditions, the students should be made aware of the fact that the adults make extensive nocturnal flights. The structure of the wings should be examined in the preserved specimen.

Most of the investigation can usefully be concentrated on the structure and functions of the first and third pairs of legs. In the males of a number of dytiscid species, the first three joints of the tarsi of the front or first pair of legs are dilated to form efficient adhesive organs (figure 13.1.1) The student can identify the cup-like suckers on these organs easily with a

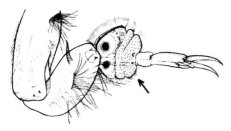

Figure 13.1.1 Front leg of male *Dytiscus marginalis*, showing the sucker pad.

hand lens. Their function can be observed in a feeding specimen or in a mating couple. The male will retain hold of a female for many hours with the aid of these adhesive pads.

The hind or third pair of legs can be seen to be large, widely spaced, flattened, and equipped with fringes of long hairs which make them extremely efficient 'oars'. Removal of the hair fringes and removal of the complete 'oar blades' demonstrate the importance of these organs to the diving beetle. The beetles normally are lighter than water and always tend to float back up to the surface. These oars are essential if the animal is to penetrate to any depth in the water.

The student should also note that although an oarless specimen now resembles a terrestial beetle floundering at the surface of the water, its pattern of leg movements is different. The back legs of *Dytiscus* beat together in unison, whereas other beetles may show the normal movements of a land beetle where the legs move alternately. The locomotion of *Dytiscus* can be compared (if desired) with that of (for example) the Great Silver Beetle *Hydrophilus piceus*, which swims rather jerkily and inefficiently by moving its legs alternately. For a detailed treatment of this topic see Rockstein (1974).

Time
1 hour

13.2 Feeding in *D. marginalis* larvae and adults

Purpose
The feeding behaviour of both adults and larvae is dramatic and immediately impressive to the student. The suggestions here are designed to exploit this easily

produced and easily observed behaviour as a start to simple open-ended investigations into food location and food preferences.

Preparation
Larvae and adults should be starved for at least 24 hours prior to their use in class.

Materials

Per student
1. 1 *Dytiscus* larva in beaker of pond water
2. 1 *Dytiscus* imago in beaker of pond water
3. 1 observation cell or small tank of pond water
4. 1 bench lamp
5. 1 × 3, or × 4, large magnifier or stereomicroscope
6. 1 stand and clamps
7. 1 pair long forceps
8. 1 earthworm (or alternative live food)
9. 2 pipettes
10. 1 watch glass

Methods
The exercise is initially one of simply observing the larva and adult finding, capturing and feeding on a food species such as an earthworm. An essential feature of the exercise is to organize a good viewing position. The observation cell or small aquarium should be small enough to keep the animal in a restricted viewing area. We suggest a plate glass or clean new Perspex unit about 20 cm × 20 cm × 5 cm in size, clamped upright so that it can be efficiently illuminated and viewed at eye level through a suitable magnifier.

The larva, if anything more voracious than the adult, will readily attack and consume the food offered by piercing the victim with its long sickle-shaped mandibles. A secretion from the midgut is injected into the victim via the fine canals which run from the pharynx down the length of the mandibles to minute holes at the piercing apices. Digestion of the prey then takes place externally. The imago likewise will readily attack, capture, and consume the offered earthworm, but in this case the digestion is internal. The larva usually feeds on the floor of the cell, while the adult, in the absence of anything to hold on to, will tend to float to the surface and feed there. The positioning of lamps and magnifiers therefore should be organized with this in mind.

Once the capture and feeding behaviour have been observed, investigations can proceed along two lines.

At a very simple level, the young student can investigate the types and sizes of animals preyed upon
successfully by both larvae and adults. From this
study, which can be spread over a number of weeks,
the student can also get a measure of the amount of
food consumed. A second approach possible is that of
trying to find out how the larvae and adult beetles
locate their prey. Tests can be devised by students to
find out if the animal hunts by sight; for example, a
food species can be presented to the predator in a fine
glass tube, or separated from the predator by a glass
sheet in the observation cell. They can further test the
animal's responses to the 'smell' of a potential prey
(i.e. chemolocation). The adult beetle seems to bump
into its prey more by accident than anything else, but
a rise in the level of hunting activity can be detected
when food is present, and this requires some thought
on the part of the student as to how this might be
quantified. Extracts of prey animals can be presented
to the predators by pipette to find out if this results
in an increase in search activity or to see if attacks on
the mouth of the pipette can be elicited. Tinbergen
(1951) found that a tadpole in glass tubing elicited no
response whereas meat extract caused the *Dytiscus* to
grab at anything it touched.

Time
1 hour or more

13.3. Transverse orientation to light in *Acilius* larvae

Purpose
Some fishes, crustaceans and water beetle larvae show
the transverse orientation to light which has been
termed the Dorsal or Ventral Light Reaction
(Fraenkel and Gunn, 1961). The Carplouse *Argulus
foliaceus,* for example, always turns the dorsal surface
toward the incident light, whether it comes from
top, bottom or sides. Likewise, the Brine Shrimp
Artemia salina orientates so that its ventral surface is
turned towards the light. The present exercise is to
demonstrate this reaction in dytiscid larvae, and
Acilius larvae show it particularly well.

Preparation
None

Materials

Per 2 students:
1. Observation cell or small aquarium, about 10 cm ×
   10 cm × 10 cm, filled with filtered pond water
2. Bench lamp or ray box
3. Stand and clamps to hold the small tank (1) so that it
   can be illuminated from below
4. Containers with three or four *Acilius* larvae
5. Small net or wide-mouthed pipette

Methods
The larvae are released, singly if desired, into the
observation cell. The bench lamp is then shone into
the tank first from above, then from below and finally
through the side. In each case the orientation of the
larva is noted. Ensure that extraneous light does not
interfere with the tests.

The larva normally turns so that its dorsal surface
is towards the light. When the small aquarium is illuminated from below, the animal will swim towards
the bottom as if it were the surface to take in air. But
the student should note that although the dorsal light
reaction is shown, the direction taken by the animal,
i.e. whether towards or away from light, is determined
by its internal state. If it requires air, it will move
toward light, afterwards it swims away from the light.
That there is no fixed reflex-like stimulus-response
relationship has been clearly demonstrated by Schöne
(1962).

Time
20 minutes

REFERENCES

Fraenkel, G. S. and Gunn, D. L. (1961), *The Orientation of
    Animals*, New York, Dover.
Rockstein, M. (1974), *The Physiology of Insecta*, 2 ed., Vol. 3:
    'Locomotion Mechanics and Hydrodynamics of Swimming in
    Aquatic Insects,' by Nachtigall.
Schöne, H. (1962), 'Optisch gesteuerte Lageänderungen (Ver-
    suche an Dytiscidenlarven)' *Z. Vergleich. Physiol.*, 45,
    590-604.
Tinbergen, N. (1951), *The Study of Instinct*, Clarendon Press, Ox-
    ford.

14. Mosquito Larvae

INTRODUCTION

Mosquitoes are Diptera of the Family Culicidae. They have aquatic larvae and pupae, but they breathe air through respiratory siphons, a single one in larvae and a pair in pupae. The larvae and pupae must therefore come to the surface of the water frequently in order to breathe. Some species have to leave the water surface in order to feed, but others can feed while still remaining in contact with the water surface. A large number of species will dive from the surface

to the substrate in response to a variety of stimuli which may indicate potential danger. The exercises described here are based on the yellow-fever mosquito *Aedes aegypti,* but apply to a great variety of larvae of the Subfamily Culicinae, where there is a distinct breathing siphon, and larvae rest at the surface with their heads hanging downwards (figure 14.0.1). In the genus *Anopheles* there is a very short siphon and larvae rest parallel to the water surface (figure 14.0.2).

Figure 14.0.2 Larva of *Anopheles* mosquito, showing very short siphon and body position parallel to and immediately below the water surface (after Gillett, 1971).

Escape responses are shown by this genus, but they may be either diving downward or sideways runs, while maintaining contact with the water surface. So the exercises described below could be conducted on *Anopheles* species, but the results would be different and more complicated.

Aedes aegypti is a tropical species so, if it is to be used in Britain or other temperate regions, it has to be kept in laboratory culture. Other species may be collected as eggs or larvae from streams or outside water tanks, and in tropical regions these may be both locally abundant and available for the greater part of the year. In Britain and other temperate regions, larvae tend not to be found in great abundance and are

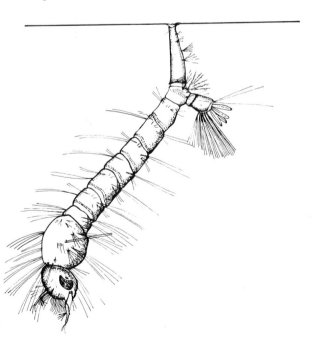

Figure 14.0.1 Larva of *Culex* mosquito, showing a distinct breathing siphon at the water surface and the body angled down into the water (after Gillett, 1971).

normally found in the summer. British species may be identified with the aid of Marshall (1938). Larvae may be obtained from laboratory suppliers, but in Britain this is an unreliable source. Laboratory maintenance of a colony of *Aedes aegypti,* or any other species, is difficult but a number of universities and other institutions do keep colonies for their research. Eggs may be begged from them, and the larvae fed at first on a fine yeast suspension, and later on wheat germ or 'Bemax'. Larvae will grow to the 4th (final) instar in 10 days or so, at 25°C.

EXPERIMENTAL WORK

14.1. Habituation of diving response

Purpose
To illustrate the diving response and the waning of that response on repeated presentation of a stimulus, i.e. habituation. To show that habituation is fairly stimulus-specific, that the response recovers rapidly, and that habituation is not muscular fatigue.

Preparation
None

Materials

Per pair of students:
1. Ten 3rd or 4th instar larvae
2. One 7.5-cm diameter crystallizing dish
3. One Pasteur pipette

Methods
The larvae should be placed in the crystallizing dish in about 3 cm of water. The larvae should then be allowed about 10 minutes in the dish to settle down. Larvae will begin to swim slowly up and down between floor and water surface, but at any one time there will be 5 or 6 larvae at the water surface. If a hand is now passed over the dish about 30 cm above it, most of those larvae at the surface will be seen to dive quite rapidly to the floor and lie there. This is the diving response. In less than a minute the larvae will be back at the surface again, and the stimulus can be repeated again. After the hand movement has been repeated about 5 times, however, the diving response is no longer

shown. The larvae have habituated to the stimulus. The short-term nature of this behaviour change can be shown by leaving the larvae for 4 to 5 minutes and passing a hand over them again. The response will be seen to have recovered.

The same diving response by larvae at the surface can be seen in response to two other kinds of stimuli, firstly dropping a drop of water from a Pasteur pipette on to the water surface, and secondly tapping the side of the dish gently with a pencil. The first stimulus produces a surface ripple on the water, the latter predominantly pressure waves within the water (vibration). Habituation rapidly occurs in response to both these stimuli.

There are two other features of habituation which are important and easy to demonstrate. These are:

(a) Habituation is not due to muscular fatigue.
(b) Habituation is fairly stimulus-specific.

Both these points may be illustrated by habituating larvae to a hand moving over the dish, and then dropping a drop of water into the dish and observing that larvae at the surface immediately dive.

Time
10-20 minutes

14.2. Investigation of diving response

Purpose
To set up an open-ended investigation so that all the points mentioned in 14.1 relating to responsiveness and habituation may be encountered by the student. In addition the experimental nature of the exercise requires the student to:

(a) Design controlled experiments
(b) Standardize stimuli
(c) Repeat the test situation several times.

Preparation
It is advisable to fill the crystallizing dishes with water about 2 hours beforehand, because small bubbles are liable to form on the sides of the dish; if not removed, these get caught in the larval mouth parts and disturb the behaviour.

Materials

Per pair of students:
1. 10 mosquito larvae
2. 7.5-cm crystallizing dishes
3. 2 retort stands with clamps
4. 2 pieces of white card 10 cm × 15 cm
5. 2 pieces of black card 10 cm × 15 cm
6. 2 plastic beads with holes through them, one 1 cm diameter, one 2 cm diameter (approx.)
7. 2 or 3 elastic bands
8. 100g Plasticine
9. 2 Pasteur pipettes
10. 1 protractor
11. 1 bench lamp
12. 1 stop clock
13. 1.5 m cotton thread
14. 1 m string

Methods

The students are asked initially to observe the response of larvae to the three types of stimuli:

(*a*) Passing a hand over the dish
(*b*) Dropping a drop of water on to the water surface
(*c*) Lightly tapping the side of the dish with a pencil.

They should be warned that excessive repetition of the stimuli may cause a loss of response, or they may find, after some preliminary 'playing around', that they cannot elicit any response when they come to make systematic observations.

The response to the three kinds of stimuli should be carefully described, and compared with one another and with the behaviour of the larvae when they are not being stimulated.

The students should see that normal swimming to and from the surface occurs all the time. It is a wriggling motion, but not a very fast one. Presentation of any one of the stimuli causes larvae at the surface to dive to the floor and stop. This is a wriggling motion, but more rapid than normal. The diving response to all three stimuli appears to be the same.

When the dish is tapped, larvae on the floor of the dish will run rapidly sideways across the floor, sometimes for several seconds after the stimulus has been applied; however, for hand movement and water drop there is no response from those larvae not at the surface. Having made preliminary observations, the students may now start experimental investigation.

To help them decide what they are going to investigate, the following questions may be presented to them:

(*a*) Does repetition of a stimulus alter the percentage response?

(*b*) What is the relationship between stimulus frequency and percentage response of larvae?
(*c*) Does repetition of one type of stimulus (e.g. movement) alter the percentage response of larvae to another type of stimulus (e.g. surface ripple)?
(*d*) What is the relationship between the percentage response of larvae and the properties of the stimulus (e.g. speed, size, intensity, etc.)?

The response to movement of a hand could be due to the movement or a change in light intensity. In fact it is mostly the change in light intensity. The stimulus can be standardized by putting the dish under a bench lamp which is then switched on and off. This produces a good response plus obvious habituation. A curious response can be noticed in *Aedes* by observant students: whereas diving in response to *light off* is immediate, to *light on* it occurs with a 1-2 second delay. The reasons for this are not known. It can also be shown that hand movements across the dish can be made so slowly or so quickly that no response occurs. A way to produce a much more standard moving stimulus is to have a disc of black or white card swinging on a piece of string from the retort clamp. The angle of swing can be controlled by attaching the protractor at the top of the string. Using discs of different diameters it will be found that discs of 1 cm diameter are almost without effect, but that the response rapidly improves as the stimulus gets bigger. The water drop can be standardized by fixing the Pasteur pipette at a specific height above the water surface. Standardization of vibration can be obtained by threading one or other of the beads on to a fixed length of string attached at the other end to a retort clamp. The string is then deflected through a standard number of degrees before being released. The sides of the dish should be covered with paper to exclude the possibility of the swinging bead being seen by the larvae.

Good habituation will be seen to all stimuli, and recovery of response to water drop will be seen clearly after habituation to movement or change in light intensity. The water drop with tap on the dish does not work quite so well, probably because they share some features in common.

Time
45 minutes

REFERENCES
Marshall, J. F. (1938), *The British Mosquitoes*, British Museum Publications.

15. Fruit Flies

INTRODUCTION

Drosophila melanogaster, the fruit fly is so well known as a laboratory animal that it needs little introduction here. *Drosophila* is a genus of small muscid dipterans with a worldwide distribution. In general, both the adults and larvae feed on yeasts and bacteria occurring in fermenting substances such as rotting fruit, plant saps and flower nectars.

D. melanogaster was introduced towards the beginning of this century as a convenient animal for the study of the genetic determination of body features such as eye colour and wing shape. Today it has become one of the most celebrated of the geneticists' experimental animals.

Recently, however, *Drosophila* has also attracted the interest of ethologists as an animal suitable for the study of the inheritance of behaviour and, for example, the behavioural mechanisms responsible for speciation.

The culture and maintenance of *Drosophila* is relatively simple, and fully documented in a large number of biology textbooks. Colonies are easily maintained in the laboratory in small vials, bottles, or large specimen tubes, at 28°C on a standard medium as described by Shorrocks (1972). Supplies of flies, larvae, eggs, tubes and media are readily available from most biological supply houses.

EXPERIMENTAL WORK

15.1. Courtship behaviour

Purpose
The courtship behaviour shown by male *Drosophila melanogaster* to females is, elaborate, and the responses of the females to prevent premature mating by the male are of several types and fairly easily recognized. A feature of male displays is prominent wing movements, which suggests that some aspect of wing movement is an important component of the courtship. This can be tested by investigating the mating success of partially winged males.

Many species of *Drosophila* may be found in the same geographical area, so a possible reason for elaborate courtship within the genus is to serve as a species-isolating mechanism, with each sympatric species having a somewhat different male courtship to enable females to repel all males but conspecifics. This means that between-species courtship should be less successful than within species; a prediction which may be tested.

Preparation
From the point of view of planning the class, it is obviously more convenient to have continuous laboratory cultures of at least one species, preferably *D. melanogaster* since this is the best known. If eggs or larvae are to be obtained from an outside source, especially for the class, it is useful to know that *D. melanogaster* and *D. simulans* grow from egg to adult in about nine days at 25°C. The adults should be timed to hatch three-four days before the class. The females are very unwilling to mate on the first day

after hatching, which allows the males and females to be separated before mating occurs, so that virgin animals are available for the class; this is necessary as mated females reject further male advances. Males and females should be separated by lightly anaesthetizing the animals with ether and individually sorting males and females into separate containers with food medium to keep them in good condition until the class.

In *D. melanogaster* the abdomen of the female gradually increases in darkness towards the posterior; the abdomen of the male has posterior abdominal segments distinctly darker than the others (figure 15.1.1). *D. simulans* resembles *D. melanogaster* very closely, and the sexes may be distinguished in the same way. The sexual receptivity of females does not develop until the second day after emergence, and is optimal on days three and four. After a week the performance of males and females will markedly decline.

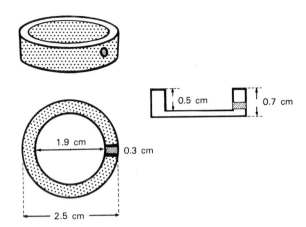

Figure 15.1.2 Perspex cell which, when covered with a cover slip, forms the cell inside which the courtship of the flies may be observed.

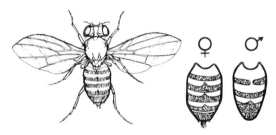

Figure 15.1.1 General view of intact female *Drosophila melanogaster*. Diagrammatic representation of the abdomen shows that of the female to be a fairly uniform dark colour, but that of the male to have a concentration of pigment towards the posterior end.

Since the observation cells for observing courtship are non-standard equipment, they will have to be made for the class. They are of Perspex and of the specification shown in figure 15.1.2. This can be made from pieces of Perspex rod cut to length and milled to produce the cell cavity.

Glass funnels to fit on top of the culture jars to introduce flies into the observation cells will also need to be made (figure 15.1.3).

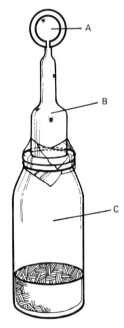

Figure 15.1.3 Attachment of a glass funnel (B) to the top of a culture bottle (C) by means of adhesive tape allows the transfer of flies into a Perspex observation cell (A), since flies tend to climb up the funnel.

Materials

Per pair of students:
1. A stereomicroscope
2. 4 Perspex observation cells (figure 15.1.2)
3. A stop watch

4. 6 coverslips to fit over the top of the observation cells (either round or square).
5. A small ball of cotton wool.

Per class:
1. A bottle of virgin female *D. melanogaster*
2. A bottle of virgin male *D. melanogaster*
3. A bottle of virgin male, 'dumpy' or 'vestigal' wing mutant *D. melanogaster*

4. A bottle of virgin female *D. simulans*
5. A bottle of virgin male *D. simulans*
6. Glass funnels, to be attached one to each of the culture bottles (15.1.3)
7. Cotton wool.

Methods

To obtain a pair of *D. melanogaster* in an observation cell, a coverslip should first be placed over the top of the cell, held on with a lick of saliva. One female fly should then be persuaded to migrate up the funnel of the culture jar and through the hole in the side of the cell (figure 15.1.3). A male should then be introduced in the same way, and the hole then sealed with a plug of damp cotton wool. The cell may be placed under the microscope and the behaviour of the flies observed.

Description of courtship: After a few moments to settle down, the male will start to take an interest in the female. He will first approach her and tap her with his front tarsi. This may be to help him check if she is the right species. The male then takes up a position close to the female and facing towards her. If she moves, he follows, keeping a fixed distance from her. This phase of courtship by the male has been called *orientation*. The male then begins to show wing display, which may be one of two kinds: *scissoring* and *vibration*. In the former both wings are scissored open and shut, opening gradually each time until they are held open and slightly raised; this position is briefly held before the wings are closed again. This movement is usually performed when the male is

standing beside the female orientated towards her. The more common wing display is, however, *vibration;* in this the male stands usually beside the female, extending one wing almost at right angles to his body and vibrating it up and down (figure 15.1.4.). The extended wing is usually the one nearest the head of the female. After some bouts of orientation and wing display, the male may approach the female from behind and show *licking* and *attempted mating*. These two movements are often performed together, the male extending his proboscis to lick the female genitalia, at the same time curling his abdomen forwards in an attempt to mate (Manning, 1960).

The responses of the female to the first advances of the male will probably not be favourable; she may demonstrate her unreceptiveness simply by *running away;* she may, however, repel the male with a *wing flick* directed towards him or by *pushing away* with one leg. If he approaches from behind the female may *extrude her genitalia* which inhibits mating in the male.

The pattern of courtship, although containing distinct elements of behaviour, does not have a predictable overall sequence of stimulus and response between male and female; the male does, however, gradually show more licking and attempted mounting towards the end of courtship and, if the female is in reproductive condition, mating should occur within five minutes.

The courtship of male *D. simulans* contains all the same elements as *D. melanogaster* but differs in that courtship is not so active, and also in that its most common wing display is scissoring, not vibrating (Manning, 1960).

Mating with wingless males: To observe this, a normal female *D. melanogaster* should be introduced into a cell with a male with reduced wings. The males should either be of the mutant 'dumpy,' which has wings reduced to two-thirds normal length, or 'vestigial' which has wings reduced to vestiges. If these are unavailable, normal males may be anaesthetized with ether, and all or part of the wings cut off.

The behaviour of these courting males will be essentially the same as that of normal animals, except that their wing displays will be reduced or absent. If the courtship duration is compared between reduced-wing males and normal ones, it will be found

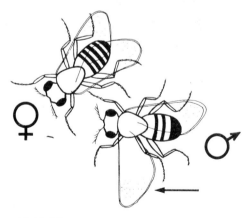

Figure 15.1.4 Diagram of the courtship position of a male fly, facing towards the female and extending and vibrating the wing nearest the female's head.

that reduction of wings prolongs courtship; that is to say, makes it less effective.

A reasonable sample for such a comparsion may be achieved by pooling class results. The reduced effectiveness of the courtship seems largely to be due to the loss of the *song* produced by the vibration of the wings. This song is a train of sound pulses with a frequency and pulse interval characteristic for the species (Bennett-Clark and Ewing, 1970). This song is detected by the arista, feathery extensions of the antennae (figure 15.1.5) of the female, which is the reason that the male normally vibrates the wing nearest the female's head. It should, however, be pointed out that the reduced effectiveness of the

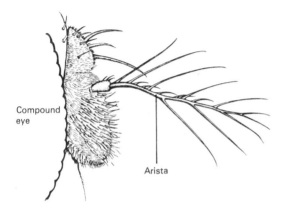

Figure 15.1.5 Detail of the antenna of female *Drosophila*, showing the stout basal segments to which is attached the long thin *arista* which detects the wing vibrations of the courting male (after Bennett-Clark and Ewing, 1970).

courtship of the male mutants may be due to other defects in their behaviour.

Courtship between species: The courtship song shown by many *Drosophila* species appear to be a method of maintaining reproductive isolation among species which are not geographically isolated. *D. melanogaster* and *D. simulans* both have a world-wide distribution, and can be persuaded to court females of the other species. A comparison of courtship duration between species will show that it takes longer than courtship between pairs of the same species.

Time
In two hours it should be possible to observe carefully and time the courtship of two normal pairs of *D. melanogaster*, the courtship of two males with reduced wings, and of two males of *D. simulans* given *D. melanogaster* females.

Note. For cleaning or preening behaviour in *Drosophila*, see Section 16.2.

REFERENCES

Bennett-Clark, H. C. and Ewing, A. W. (1970), 'The Love Song of the Fruit Fly', *Scient. Am.*, 223, (1): 84-92.
Manning, A. (1960), 'The Sexual Behaviour of Two Sibling *Drosophila* Species', *Behaviour*, 15: 123-145.
Shorrocks, B. (1972), *Invertebrate Types: Drosophila*, Ginn & Company, Ltd., London.

16. Blowflies and Larvae

INTRODUCTION

The genus *Calliphora*, which includes the common blowfly or bluebottle, is cosmopolitan. The adult flies are commonly found around human habitations. The males rarely enter buildings and feed exclusively on nectar. The females, however, also feed on the juices on the surface of carrion, and normally lay their eggs on such decaying substances. Each female may lay up to a thousand eggs, which begin to be deposited, in groups of up to about a hundred, some three to four weeks after emergence from the pupa. The male on the other hand is fully developed sexually within a few hours of emergence. As soon as they hatch, the larvae burrow into the decaying flesh, and feed and grow for about a fortnight. In cooler conditions the life cycle takes longer. The fully grown maggots then leave the carrion and burrow into the ground, where they pupate. After about two weeks in summer (or several months in a cold winter) the imago emerges.

Blowflies are easily kept in the laboratory, and supplies of the larvae are readily obtainable from Fishing Tackle shops, the biological supply houses, and local sources. There are a number of ways of housing and culturing the flies. Whatever container is used, however, it will be important to pay attention to the following points.

(a) The container must be washable.
(b) The container must be escape-proof and yet allow easy access to the flies and their larvae.

Thus it is suggested that a 60 cm × 40 cm × 30 cm aquarium tank suitably modified, makes a simple robust escape-proof and readily-washable house for the flies. A possible modification is shown in figure

16.0.1. The flies and larvae can be handled via a nylon net sleeve, or by way of a gauze door after anaesthetizing the flies with CO_2 from a cylinder.

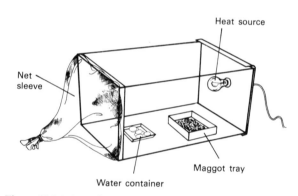

Figure 16.0.1 Aquarium modified as breeding tank for blowflies. Note water supply guarded by filter paper and deep tray of sawdust and raw meat for larvae.

EXPERIMENTAL WORK

16.1. Heat preferences of blowflies

Purpose

Blowflies are found active around human habitations whenever the temperature rises above 12°C. On cool days blowflies can be observed sunning themselves on the trunks of trees and light-coloured walls which are facing the sun. Even in the laboratory tank they can be seen to prefer a warm site near the light bulb. This simple exercise is merely to demonstrate that the flies do have a temperature preference and set the scene for a fuller analysis of this behaviour.

Preparation
None

Materials

Per group of students:
1. 20 blowflies in a heat preference chamber.
2. CO_2 cylinder
3. Paint brush

Methods
The heat preference chamber is simply an aquarium tank or Perspex chamber modified as in figure 16.1.1. Essentially it is a chamber with a 40-watt light bulb almost touching the centre of one of the walls. This wall should be at least 30 cm × 30 cm square. When the bulb is lighted, the wall will become fairly hot at a centre point, and progressively cooler towards the edges. Thus concentric zones of temperature will be set up. The flies will tend to settle in a particular zone around the very warm central area.

Figure 16.1.1 The principal components of a simple temperature preference chamber.

If the main stock tank has not been washed for a while, it is possible to see this zone marked out quite distinctly around the light bulb by the droppings of the flies. Indeed this has usually been our initial observation which has led to the experimental work.

Once the existence of a preferred zone round the heat source has been established, it is possible to test hypotheses such as the following:

(a) The distribution of the flies is determined by heat and not light.
(b) The distribution of the flies is determined by heat and not the configuration of the interior of the chamber.

In (*a*) the problem will be to set up a concentric heat gradient or pattern in the wall of the chamber, with an opaque barrier between bulb and wall. The temperatures of different zones can be measured by thermometer or thermocouple, and sketched in with a felt pen. In (*b*) students who are worried about the shape and size of the warm surface may simply move the lamp so that a different pattern of heat zones is established; the flies will usually land and settle more frequently in the new position of the preferred heat zone.

The chief problem with this piece of work is screening the flies from extraneous vibration and movements which cause the animals to scatter in the chamber. The flies must be given the opportunity to settle completely undisturbed, and yet must be open to observation. Black paper screens with peepholes were set up by some students, and these proved highly successful.

Time
Up to 1 hour for the initial observations.

16.2. Cleaning or preening behaviour

Purpose
The repertoire of cleaning or preening behaviour of blowflies is a delight to watch because of the apparent stereotypy of the individual movements, coupled with the variety of components in the whole repertoire. Complex preening movements are very widespread among the Diptera. Adult mosquitoes, for example, show them after feeding, so an exercise similar to the one described here could be tried out on many species. The exercise described below may certainly be carried out on the genera *Drosophila*, *Sarcophaga* and *Musca* with similar results to those for *Calliphora*. There is scope for a comparative study between, say, *Calliphora* and *Drosophila* since they have many components of the repertoire in common, but two or three which are not shared.

It is surprising that fly preening has not been more studied, but the behaviour of *Drosophila melanogaster* has been described by Szebenyi (1969). This study exemplifies one type of exercise which may be attempted and is purely descriptive; the purpose of the exercise is to observe carefully a complex sequence, to divide it up into components, and to describe each one. At a more advanced level, a study

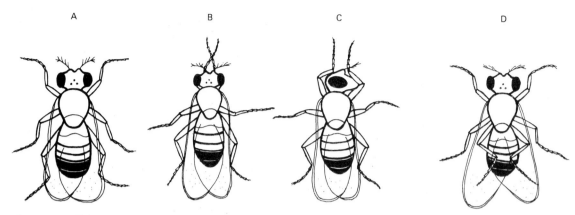

A   B   C   D

can be made of the sequences of components within the repertoire in order to establish the nature of the mechanism which generates the repertoire. The mechanism may further be investigated by experimental manipulation of the animal or its environment.

The great merit of this exercise is that the basic situation is very simple (a fly in a box) and this can be used to undertake an exercise whose aims range from the very simple to the highly elaborate.

Preparation
None

Materials

> Per pair of students:
> 1. About six flies (*Calliphora*)
> 2. Enough 5-cm diameter pill boxes to have one fly to a box
> 3. A magnifying lens (not essential)
> 4. A tape cassette recorder

(Note on *Drosophila*: If observations are made on *Drosophila*, a small Perspex cell and stereomicroscope will be needed as described in Section 15.1.)

> Per class:
> 1. A small carbon dioxide cylinder
> 2. Sucrose
> 3. A fine paint brush

Methods

(a) Description of the components of the preening repertoire

For this part of the exercise it is not necessary to have the cassette recorder. All that is required is that the fly

Figure 16.2.1 Fly preening movements:
A – Fly in resting position with all feet on the ground.
B – Fly rubbing front legs together. C – Fly rubbing front legs on head. D – Fly rubbing hind legs under wings. Movements shown here are for *Drosophila* but are essentially the same for *Calliphora* and other genera (after Szebenyi, 1969).

should be observed and, after a period of observation during which full notes are taken, a list of the behaviour components made. As was pointed out in Part A1 (page 9), the nature of such a list depends on the level of the system we wish to describe. Szebenyi concluded that, at the lowest level of analysis, all the behaviour shown could be divided into two types of movement: *rubbing* (i.e. the rubbing of one leg against another) and *sweeping* (e.g. a wiping movement of both legs across the under surface of a wing) (figure 16.2.1). At a higher level these two movements involve different combinations of legs and parts of the body to perform several quite distinct behaviour components, e.g. front legs together (B); front legs on head (C); hind legs across under surface of wings (D). These components are fitted together to form a preening 'bout' which is defined as a group of behaviour components which occur together and are separated by some arbitrary period of time from the preceding and subsequent bout.

In this exercise it is most rewarding to concentrate on the description of the preening components, because there are enough to make careful observation necessary, but not too many to make the exercise impossible. Szebenyi (1969) recognized in *Drosophila* 7 leg-cleaning and 20 body-cleaning movements, but our impression for *Calliphora* is that some of these are either very rare or absent, and so a rather smaller list should be expected. These will include:

(a) Prothoracic legs together
(b) Prothoracic legs on head
(c) Prothoracic legs plus one mesothoracic leg (left or right)
(d) Metathoracic legs together
(e) Metathoracic legs under abdomen
(f) Metathoracic legs over abdomen
(g) Metathoracic legs under wings
(h) Metathoracic legs plus one mesothoracic leg.

(b) Sequences of preening components

As was mentioned above, a preening bout consists of a sequence of preening components of the sort listed above. This leads on to the question as to whether these components are organized with some kind of sequence within the bout or occur randomly. Students should record the sequence in which the components occur either directly on to a data sheet, or initially on to tape and then on to a data sheet. Observation will show that preening bouts are very variable in length, some consisting of one component of the repertoire only, and others of many components in sequence. A long preening sequence usually starts at the front end of the animal and proceeds to the posterior; a short preening bout therefore usually only involves the front end of the animal. Examples of these anterior preening components are (a), (b) and (c) listed above. Preening the anterior of the body usually starts and ends with prothoracic legs together. The shortest bout possible therefore is probably:

Front legs together ———→ stop.

A longer bout might be:

Front legs together ——→ front legs together ——→
front legs on head ——→ front legs together ——→
front legs on head ——→ front legs together ——→
front legs and left mesothoracic leg together ——→ front
legs together ——→stop.

Such a sequence might carry over into a sequence of posterior end preening; these sequences usually start and end with metathoracic legs together, so the shortest possible sequence involving the posterior end would probably be a single back-legs-together.

(c) Effects of making the animal dirty

The most obvious functional explanation of preening is that it is to clean the body surface of the animal. Indeed the sequence:

Front legs together ——→front legs on head ——→
front legs together could be explained as:

Cleaning the hands prior to wiping the face ——→
wiping the face clean with the hands ——→
cleaning the hands after wiping the face.

If students hypothesize this, they should predict that, in a dirtier environment, flies should preen more. The sprinkling of a bit of chalk dust in the box will show this to be the case.

Another hypothesis which is also testable is that a substance sticking to a particular part of the body will lead to an increase in the component concerned with preening that part, not an increase in all components. This may be easily effected by anaesthetizing the fly with a whiff of carbon dioxide and a spot of concentrated sugar solution placed on one wing with a paint brush. It is better to put the spot on the posterior end rather than, say, the head, because a big increase in a relatively uncommon component like leg-over-wings would be more obvious than an increase in head preening (which is frequent anyway). We do not have a great deal of data on this situation, but class results suggest that the effect of enhancing particular components of the sequence is not marked; there is plenty of scope here for applying different substances to different parts of the body. There is, however, one problem worth mentioning which may be obscuring a real change in preening resulting from local application of some substance; it is that the carbon dioxide anaesthesia by itself tends to enhance preening. This general increase may serve to mask the effect of the experimental procedure; it emphasizes the need for a control where the flies have been anaesthetized only. One class observation which is suggestive of an effect of localized dirt on particular preening components was that chalk dirt on the floor of the container caused a drop in the proportion of wing preening movements — the suggestion being that the preening priority had shifted towards leg-cleaning movements at the expense of wing preening.

(d) The effect of amputating legs and wings

The increase in preening due to dust demonstrates an influence of external stimuli on the preening; the fairly rigid sequences of components suggest a fairly fixed centrally generated pattern. This raises the question of the extent of external stimulation on the central pattern. If, for example, one front leg is amputated, will the remaining leg go through the motions of front-legs-together in the absence of contact stimulation from the missing leg?

This question can be very quickly resolved by anaesthetizing a fly with CO_2, cutting off one prothoracic leg as far down as possible, and observing the animal when it comes round. Almost immediately the fly will be seen to perform front-legs-together with the remaining leg alone. Very similar observations may be obtained by removing one back leg, or one or both wings. The initial observation is important in itself, but it may be that this is not the whole story. Fairly casual observations suggest that there is, after ten or more minutes, an increase in the occurrence of the remaining front leg being rubbed against the mesothoracic leg on the same side; similarly for a single remaining metathoracic leg against the mesothoracic on the same side. This modification of the preening behaviour due to loss of limbs could therefore prove to be an interesting project.

Time

To observe and classify the behaviour repertoire as suggested in Part *a* will probably take at least 40 minutes. Part *b* is not really suited to a practical of 3 hours or less, because the scoring of the sequences of behaviour need practice; it is therefore best suited to a project study of 10 or more hours. The effect of making an animal dirty and increasing the amount of preening as suggested in Part *c* could be observed in a quantitative way, or shown in a demonstration in 10-15 minutes but does allow an experimental investigation of 1 or 2 hours. The immediate effects of amptuation (Part *d*) similarly can be demonstrated in about 5 minutes, but the long-term effects must be tackled in a project extending over 2 weeks or more (see Section 8.2 (page 69) on the cockroach antennal preening).

16.3. Stimuli influencing burrowing response of maggots

Purpose

Blowfly maggots are extremely easy animals to handle, and their burrowing response is easily elicited. This makes them ideal animals for a simple open-ended experimental study when limited time is available. The burrowing of the animal down into the substrate raises questions about the stimuli which *orient* the animal down into the substrate, and also about the stimuli which initiate or *release* burrowing behaviour. The fact that orienting and releasing stimuli may be

different is a problem which is likely to arise during the exercise or in discussion. Another important point about the exercise is that it can be used to break down too rigid ideas about the nature of orientation responses. There are dangers at an introductory level of the use of words like 'positive geotaxis'. Among these are that to label a response in this way may sound like a complete explanation which may inhibit further investigation; also that an orientation response may be thought of as something immutable; so, for example, a maggot having burrowed down could not 'decide' to burrow back up again.

This exercise will show students that maggots respond to stimuli; that they can respond to directional stimuli directionally (orient their responses), but that care should be taken not to give an over-simple explanation of the response observed.

Preparation

None

Materials

> Per pair of students:
> 1. 20 maggots
> 2. Four 7-cm diameter crystallizing dishes
> 3. Two 12-cm diameter crystallizing dishes
> 4. 1 bench lamp
> 5. One 400-ml beaker
> 6. Bunsen burner and tripod
> 7. 1 square of black paper (30 cm × 30 cm)
> 8. A mercury thermometer
>
> Per class:
> 1. A bucket of sawdust
> 2. A bucket of peat
> 3. A roll of adhesive tape
> 4. A graded series of sieves

Methods

A single maggot should be placed on the bench and allowed to crawl. When the maggot is progressing in a particular direction, a bench lamp should be placed in front of it and turned on. The maggot will respond by turning away from the light. This should be repeated two or three times.

About five maggots should then be placed on top of a layer of peat in a 7-cm diameter crystallizing dish with a bench lamp placed immediately above it. The maggots quickly burrow into the peat. Students should then be asked to repeat the experiment, except that as soon as the maggots are placed on the peat, the dish should be placed in a dark cupboard. Students should be clear what hypothesis this will test, and should be asked to predict the result of the experiment. Many students predict that the maggots will not burrow, having already observed that maggots move away from light. On opening the cupboard after five minutes, some of the maggots will be found to have burrowed. This simple exercise may be used to form the basis of a discussion or the start of an open-ended investigation.

A great variety of experiments are possible in an open-ended exercise, but the following lines of investigation usually develop: the influence of light on burrowing may be investigated by observing how many maggots burrow into the substrate within a given time when dishes are kept either in the light or the dark. The effect of the light is to speed up burrowing slightly, and therefore has the effect of at least enhancing the burrowing behaviour.

Shining a light from underneath the dish causes the larvae to avoid the bottom; however, this may be partly due to the heating of the glass. Heat can be shown to be a factor influencing burrowing since, if a dish containing peat with maggots in it is floated in a larger dish of warm water, the maggots come to the surface again.

A comparison may be made between burrowing in sawdust and peat. This shows that the time taken to disappear into the peat is less than for sawdust. In a choice test of half a dish of peat and half of sawdust, more maggots burrow into the peat side, and those burrowing in the sawdust may come up to the surface again. Students trying to explain the differences between maggots in the two substrates realize that there may be any number of differences between them. The easiest of these to investigate are humidity and texture. Humidity can be investigated by timing burrowing into dry and damp sawdust. Texture may be investigated by separating different particle sizes of peat or sawdust and observing the burrowing rate in them. However, another problem is raised by these experiments. This is that a difference in burrowing time between different substrates may indicate a real preference of the larvae, or it may be that one substrate is easier to burrow in than another.

Time
The initial exercise may be completed in as little as ten

minutes, to provide the basis of a short discussion. A useful open-ended investigation can be achieved in two hours but, because the class results will be various and possibly also contradictory, it is essential to provide discussion time in addition.

16.4. Influence of pupation site on orientation of pupae and pupation delay

Purpose
Larvae of *Calliphora* and *Sarcophaga*, when they are fully grown, stop feeding, move away from the food source, and search for a dry crevice where they pupate three to five days after the beginning of this migration. This exercise deals with three aspects of the pupation behaviour.

  (a) The nature of pupation sites
  (b) The orientation of pupae in the pupation sites
  (c) The pupation delay shown by larvae in unsuitable pupation sites.

Preparation
The larvae required for this exercise should have stopped feeding and commenced migration. If a laboratory colony is being kept, then this stage may be identified by the maggots moving away from the food. If maggots are being obtained from a supplier where they are already separated from the food, then maggots approaching pupation may be identified by the absence of any crop contents, whereas larvae which would still be feeding usually retain crop contents which may be seen as a dark patch towards the anterior of the larvae. Prior to every experiment, therefore, migrating larvae only should be selected from stock.

Materials

  1. At least thirty 20-cm lengths of glass or polythene tube with 5-mm internal diameter
  2. Two or three plastic trays, about 35 m × 30 m × 4 cm
  3. Some fine cotton or wire net
  4. 500g Polyfilla
  5. Roll of Sellotape

Methods
For the first experiment, the apparatus required is a plastic tray with one end filled with 20-cm lengths of

tube placed side by side with one end hard against the end wall of the tray (figure 16.4.1). The walls of the tray should be unclimbable by maggots. The floor of the open end of the tray should be raised to the level of the entrance of the tubes by being lined with Polyfilla 1-2 mm thick. Into the open end of the tray should be placed about as many maggots as there are tubes. The tray should then be placed in complete darkness for 4-5 days before the position and orientation of the pupae is recorded. It will be found that almost all the pupae are in the tubes, and that they are oriented with their heads facing the open end of the tube.

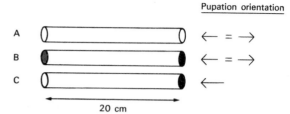

Pupation orientation

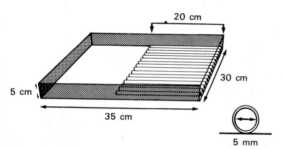

Figure 16.4.2 The 3 different experimental situations and the effect on the orientation of larvae at pupation.
A – open at both ends. B – blocked at one end and covered with net at the other. C – blocked at one end only (after Zanforlin, 1969).

if a maggot falls out of a tube it cannot re-enter. One maggot should be placed in each tube, and the experiment left for five days in the dark.

The results should be as shown in figure 16.4.2. In tubes with both ends open, pupae face equally in both directions contrary to the prediction of hypothesis (1).

In the tube with one end closed with net and the other with Polyfilla, in which it is presumed that an oxygen gradient has been established, pupae still face equally in both directions, contrary to the prediction of hypothesis (2). However, in tubes with one end closed, pupae face towards the open end, irrespective of whether the larvae are introduced at the open or closed end. The prediction of hypothesis (3) is therefore confirmed. This experiment was originally conducted by Zanforlin (1969) with *Sarcophaga barbata*, but it can also be demonstrated in *Calliphora*.

Zanforlin (1969) reported an additional feature of the experiment which was that larvae in tubes closed at both ends took longer to pupate (9.5 days) than ones in tubes with at least one open end (4 days). We were unable to confirm this in an undergraduate project with *Calliphora* maggots obtained from a supplier; however, this may well be because it is impossible to tell with supplied maggots how many days they have already been migrating. Therefore, if you wish to have students studying the pupation delay, it is probably necessary to have your own laboratory cultures of maggots. Zanforlin did extend his study of the pupation delay with some further simple experiments to test between the hypotheses:

(a) Pupation delay is caused by higher CO_2 levels in closed tubes. (Takaoka (1960) demonstrated this effect in *Drosophila melanogaster*).
(b) Pupation delay is caused by pupae establishing that the tube is closed by feeling all round it.

Figure 16.4.1 Apparatus to allow maggots to pupate in a choice of tubes. The open arena on the left leads into a row of tubes, with 5 mm internal diameter, closed at their right-hand ends.

This experiment raises the question of what stimuli cause larvae to orient towards the open end of the tube. Three hypotheses can be formulated to explain the response.

(1) When larvae detect the all-round stimulation or the sides of the tube, they turn round to face the direction from which they entered.
(2) Larvae orient in an oxygen gradient, which becomes established in the tube, so as to face the direction of higher oxygen level.
(3) Larvae respond to the closed end of the tube by pupating facing away from it.

These hypotheses may be tested by placing maggots in tubes of the three kinds shown in figure 16.4.2.

(a) Tubes open at both ends.
(b) Tubes closed at one end with Polyfilla and the other with net.
(c) Tubes closed at one end with Polyfilla

The tubes should be of the same type as in the first experiment, but raised slightly off the ground, so that

He tested between these hypotheses with the apparatus shown in figure 16.4.3 with the results shown. When larvae could feel that the tube was closed, pupation was delayed; but where they could not reach the roof of the tube, they treated it as open and pupated in the normal time. The CO_2 levels in the two kinds of tubes were assumed to be the same. Zanforlin did further experiments on the size of container and pupation delay which would be simple and interesting to repeat for *Sarcophaga,* or try out on *Calliphora,* or any other species ready to hand.

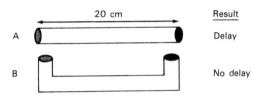

Figure 16.4.3 Experiment to study pupation delay. Larvae in situation A could feel the tube was closed and delayed pupation. In situation B where they could not feel the roof at the ends, larvae pupated normally (after Zanforlin, 1969).

One further line of investigation is worth mentioning. This concerns the apparatus shown in figure 16.4.1. In an undergraduate study there were strong indications that if as many *Calliphora* larvae were placed in the open area as there were tubes, then they pupated one to a tube, not clustered in the edge tubes

as would be expected due to the effect of the side wall of the tray. This suggests that maggots detect and respond to the presence of a 'resident' larva or pupa. At first sight this suggests the altruistic behaviour of abandoning a pupation site so as not to block the exit of another individual; however, further experiments might reveal that pupae blocking the path of an emerging adult are damaged themselves. It would, therefore, be interesting and valuable if this effect could be confirmed and studied further.

Time
This is essentially a project study. The experiments are simple to set up, but there is then a delay of about a week before the results of the experiment can be seen. So about four hours a week for at least five weeks would be a useful time allocation. With several experiments being run at the same time, the number of weeks could be lessened, but there is then the danger of carrying out experiments which subsequently prove to be useless.

REFERENCES

Szebenyi, A. L. (1969), 'Cleaning Behaviour in *Drosophila melanogaster,' Anim. Behav.,* 17, 641-651.
Takaoka, M. (1960), 'Studies of the Metamorphosis in Insects. IV. Inhibition of Pupation by Carbon Dioxide in the Mature Larva of *Drosophila melanogaster,' Embryologia,* 5, 78-84.
Zanforlin, M. (1969), 'Perception of Spatial Relationships and Pupation Delay in Fly Larvae (*Sarcophaga barbata*)," *Anim. Behav.,* 17, 323-329.

17. Periwinkles

INTRODUCTION

Littorina littorea is the largest, commonest and most widespread of the four species of periwinkle commonly found between tide marks on British shores. Wherever the tutor has access to the sea, he has access to a convenient experimental animal. Since they are abundant almost everywhere, large numbers for class work can be collected very quickly; and in the laboratory they are robust, harmless, slow-moving, easily handled, easily maintained for short periods, and generally ideal experimental animals for simple behavioural studies.

L. littorea is a highly successful littoral prosobranch; it ranges vertically from high-water neap-tide levels to mean low-water spring-tide levels, and tolerates a wide range of substrate, salinity and exposure conditions. These periwinkles are equally at home on bare rock, shingle and mud, amongst weed, even sand. They are found on shores exposed to the open sea and in the reduced salinities of sheltered estauries. Thus, maintenance conditions for this species are not critical. They are easily transported and stored for short periods in net bags or damp sacking. In the laboratory, however, they are best kept slightly damp on a covered tray in a cool refrigerator, particularly if they have to be kept for a number of days. The best behavioural results are obtained from specimens freshly collected.

EXPERIMENTAL WORK

17.1. Simple observation of mode of locomotion in *L. littorea*

Purpose
This is a simple introductory exercise in handling and observation.

Preparation
None

Materials

Per student:
1. Small beaker of seawater
2. Crystallizing dish with two or three winkles
3. Sheet of glass, 25 cm × 25 cm
4. Bench lamp
5. Hand lens
6. Tray of seawater 30 cm × 30 cm × 5 cm.

Per group of students:
1. Jar of talc powder
2. 2 paint brushes (fine)
3. Small quantity of *Ulva* or *Enteromorpha*
4. Pestle and mortar, or access to a blender (for producing seaweed extract)
5. Small container of flour.
6. Plasticine.

Methods

The winkles should be wetted with seawater. If necessary, submerge the animals in a beaker of seawater for a minute or two until they become active. Actively moving specimens can then be transferred to the glass plate. The periwinkles will normally move about freely on the glass plate, which can be turned over so that the foot of the moving specimen can be examined closely. The glass should be carefully held (or propped up with blobs of Plasticine) so that the winkles are submerged in the water in the tray; but the surface of the observation glass is reasonably dry, so that a clear view of the animal is possible.

As with the garden snail (Section 18) it is possible to develop this convenient handling and viewing technique and use it for studying feeding patterns. The plate can be very finely coated in flour paste or painted with seaweed extract, and the mode of grazing observed. The tracks of the periwinkle can then be 'developed' using the powdered-talc technique (i.e. dust plate with talc and gently swill off excess).

Time

15-20 minutes for initial observations.

17.2. Gravity responses

Purpose

To demonstrate the marked negative geotaxis shown when periwinkles are submerged in seawater following a period out of water.

Preparation

The periwinkles should be kept dry for an hour or so before the demonstration.

Materials

Per demonstrator or per group of students:
1. Tank of seawater, at least 30 cm deep
2. Sheet of glass or Perspex 25 cm × 25 cm
3. 10 periwinkles in container
4. 1 pair long forceps
5. Piece of Plasticine
6. Finely powdered talc
7. Blank paper shield for tank.

Methods

The demonstration is relatively simple. One periwinkle should be wetted with seawater and, when it becomes active, should be placed in the centre of the sheet of glass. The glass with the moving periwinkle should be immersed slowly and carefully in the tank of water. The glass can be either clamped vertically or held in position by a piece of Plasticine; and the tank shielded from light with black paper. The periwinkle now has the opportunity to crawl up, down, or at random across the glass sheet. Normally, one gets a marked upward movement immediately. The sheet of glass can then be carefully turned through 90° or 180°, and the path of the animal noted. Again the response is usually clear-cut, with the animal turning and moving straight up the glass. If required, the paths can be talced for display and discussion.

A second animal should then be tested and the paths compared. And further animals can be checked likewise if desired. This makes a simple and clear-cut demonstration of what has been termed a negative geotaxis (Fraenkel and Gunn, 1961).

The only slight difficulty which may arise is at the point of putting the moving animal on to the centre of the glass sheet. Long forceps can be used to replace a specimen if it falls off the glass before its path becomes clear.

Time

15 minutes for initial test

17.3. Light responses

Purpose

An interesting analysis of the light and gravity responses of a related species, *Littorina neritoides,* is discussed in Fraenkel and Gunn (1961). We have found it unsatisfactory as a laboratory exercise despite its value in discussion. *L. littorea* is larger, more readily available, and shows an easily demonstrated positive response to light.

Preparation

The periwinkles should be stored dry for an hour before the tests.

Materials

Per group of students:
1. Perspex tray 40 cm × 30 cm × 5 cm, full of seawater.

2. Light source, e.g. lamp or ray box
3. 10-15 periwinkles

Methods

The tray of seawater must be shielded from extraneous light. The lamp or ray box should be arranged so that it shines through an end wall of the Perspex tray, and the animals should be placed at random in the centre of the tray. Within a few minutes the periwinkles will have become active, and normally show a definite path towards the light, often ending up in clumps on the wall of the tray at the light source. This response can be shown to weaken with time, and after 20 minutes some specimens show a distinct reversal of the response, moving directly away from the light source.

Time

30 minutes

17.4. Triggering of the negative geotactic response

Purpose

In 17.1 and 17.2 immersion in seawater triggered off activity in *Littorina*. Wetting with tap water has no effect. Likewise, it will be appreciated that rainwater does not trigger activity when the animals are exposed at low tide. This exercise is designed to explore this salinity trigger and provide the opportunity for further open-ended investigation.

Preparation

The periwinkles should be stored dry for at least an hour beforehand.

Materials

Per student:
1. 20-30 periwinkles in dry beaker
2. 1 litre seawater
3. 1 litre distilled water
4. 10 small crystallizing dishes
5. 2 measuring cylinders
6. Plasticine
7. Pair of scissors
8. 3 drinking straws
9. Graph paper
10. Sheet of stiff card

Methods

The initial test is simply to put a number of periwinkles in a container of seawater, and an equal number in a container of distilled water. After a few minutes the animals in seawater will start to move, while those in the distilled water remain quiescent. (The inactive specimens in the distilled water should be left for some time to convince the student, and then can be transferred to seawater in which they will become active, thereby demonstrating that they are in fact alive.)

Keeping in mind the varied salinities likely to be encountered by this far-ranging species (from the full salinity of exposed shores to the reduced salinities of sheltered estuaries and contaminated pools) the next step in the exercise is to establish at what salinity the animals become active. Solutions of 10%, 20%, etc., seawater can quickly be prepared and the dry periwinkles tested in them.

The periwinkles' responses to reduced salinity are variable. Some will become active, some will remain quiescent, and others will become inactive after a brief period of activity. It is difficult to arrive at a clear-cut concentration (say 50% seawater in distilled water) where all animals become active.

Several open-ended approaches are now possible:

(1) A comparison of responses of periwinkles gathered from, e.g. a sheltered low-salinity area such as an estuary, with those collected from an open shore.
(2) A more chemical approach to the problem is possible. Solutions with the same colligative properties as seawater can be prepared to eliminate the possibility that the periwinkles are responding osmotically to seawater. The student can then experiment with solutions to establish the importance of the chloride ion in triggering activity.
(3) The identification of activity is itself an interesting experimental problem. Limpets when irrigated with seawater in a Petri dish raise their shells, and this movement can be magnified by attaching a light lever to the top of the shell. A similar approach works with periwinkles. At concentration of seawater where movement is difficult to detect, drinking-straw levers can be attached to the top of the shell, and the other end of the straw arranged so that it can move against a graph-paper background. The amount of movement and its duration can then be investigated.

Time

15 minutes upwards

REFERENCES

Fraenkel, G. S. and Gunn, D. L. (1961), *The Orientation of Animals*, New York: Dover.

18. Garden Snails

INTRODUCTION

Many of the larger terrestrial pulmonates are ideal animals for simple introductory exercises in behaviour. Their behavioural repertoires are admittedly somewhat limited, but they have a number of features which make them particularly suitable for beginners: they are a convenient size for handling and observing; they are slow-moving, therefore they can be used confidently on the open bench without fear of escape; they are harmless; they are easily housed and maintained for long periods; they are not 'shy' and will behave under crowded and noisy laboratory conditions – an important consideration in many school situations.

Certainly in the school, the clean, dry and often attractively marked shell makes the snail easier to handle than the slug. The simple studies outlined here are based on a common garden snail *Helix aspersa,* but other species have been used successfully over a wide range of pupils and students.

H. aspersa are normally inactive during daylight and under dry conditions, and move and feed usually at night or after rain. They feed on living vegetable matter, from which fragments are abraded by means of a toothed radula. In winter and in prolonged dry spells, they seal up the aperture of the shell with a film of mucus hardened with calcium phosphate. Like nearly all terrestrial pulmonates, they can easily be housed and maintained in glass or earthenware containers with a gauze or nylon net lid. Soil, leaves, stones, etc., are not essential. But it is important to keep the container in a cool place in reduced light. Twice a week the snails should be fed on lettuce leaves and oats which have been lightly powdered with calcium carbonate or crushed bone. Prior to feeding, the container and the snails should be lightly sprayed with clean water and the excess drained off. This keeps the container sweet, and the snails clean and in good condition.

Virtually no preparation is required other than avoiding feeding the snails for a day or two prior to the laboratory sessions. Many, if not all, snails will be quiescent at the start of a practical session, and it is essential to 'wake them up' and have them moving about actively immediately before use. This is best done by immersing the animals in tepid water for a few minutes. As soon as a snail shows signs of activity, it should be removed from the water, dabbed dry, and introduced to the experimental set-up.

EXPERIMENTAL WORK

18.1. Observation of locomotion

Purpose
To see the foot of the snail in action.

Preparation
None

Materials

Per student:
1. Sheet of glass 25 cm × 25 cm
2. ×5 lens
3. Specimen in container
4. Clamp and stand.

Methods

Put an actively moving snail on the centre of the sheet of glass. When it has taken a firm hold of the glass surface and is moving with the foot fully extended, gently and slowly turn the glass over and clamp it firmly in a horizontal position. With the aid of a low-power lens examine the foot of the snail as it moves across the undersurface of the glass. This is a very simple exercise which has been used with school pupils, as well as more mature students, the treatment and context obviously differing at these levels. The snail normally moves freely on the undersurface of the horizontal sheet of glass, even under the relatively intense light from a bench lamp, so that it is possible to examine the ripples of muscular contraction passing along the foot with ease. The exercise may, of course, be elaborated in a variety of ways by measuring the speed of the animal and seeing if this correlates with number of ripples in, or rate of passage of ripples along, the foot. The influence of environmental factors such as temperature on rate of movement may also be examined.

Time

15-20 minutes

18.2. Habituation to vibration

Purpose

To introduce the student to experimental study of the pattern of responses of a snail to vibration or mechanical shock.

Preparation

None

Materials

Per student:
1. Wooden board
2. 250-g weight
3. Ruler
4. Specimen in container
5. Stop clock

Methods

Put a snail on the middle of the smooth wooden board and wait until it is fully extended and moving freely. The board need not be clamped, but should be in-

sulated from extraneous vibrations and shocks as far as possible. Drop the weight from a small known height on to the board at a known distance from the moving animal. Repeat a number of times at regular intervals and record the responses of the animal.

There are a number of ways of illustrating habituation to vibration shock by snails. To provide a convincing demonstration at school level it is important to give a standard vibration which is large enough for the snail to retract partially, but neither so large that the snail remains inactive for a long time nor so weak that the snail does not respond to the first stimulus. Always let the snail extend fully before the next stimulus is given. The results may be presented as a descriptive account of the diminishing amount and type of response elements (stop moving, whole body retracted, tentacles retracted) or quantitatively. At more-advanced levels the degree of response of the animal can be placed on an intensity scale (say: whole body retracted = 10; tentacles only retracted = 2) and the response then plotted against stimulus intensity, frequency, or the recovery time from a standard stimulus.

Time

15 minutes to 1 hour

18.3. Responses to tentacular, dorsal and lateral contact

Purpose

To introduce the student to experimental problems associated with attempts to study the responses to touch.

Preparation

None

Materials

Per student:
1. Wooden board
2. Specimen in container
3. Probe, spatula or bristle.

Methods

When a snail is moving freely on a horizontal smooth wooden surface, touch the snail gently with a blunt probe, spatula, or bristle. The pattern of stimulation and the parts of the body stimulated can be varied. For example, a light touch every five seconds on the middle of the right lateral surface can be compared with repeated light touches on the right posterior tentacle.

The variations possible in this type of investigation depend on the level at which the teacher is presenting it. At lower levels the simple procedure outlined above may involve two problems:

(a) What constitutes a 'standard touch'? How can the stimulus intensity be controlled? How, and for what reason, should a particular regime of stimulation be planned? In other words, in tackling how the snail reacts to contact in its environment the student is confronted with basic experimental design problems.

(b) Description and interpretation of the different responses of the animal. Repeated light touching of one posterior tentacle results in the animal slowly turning towards the stimulus, not away from it as many of the young experimenters predict. In the hands of a skilled teacher this makes an excellent demonstration. With the right touch regime, the snail will turn and lift its anterior end and position itself as if to climb on to the end of the probe. If the probe is removed, the snail will still apparently attempt to climb on to something which is not there. This is, of course, an ideal time to examine the meaning and value and use of such terms as 'seeing', 'feeling', 'sensing', and so on. At higher levels, obviously, the way is opened to methods of location and analysis of function of receptors in the snail.

Time
20-90 minutes

18.4. Feeding behaviour

Purpose
To study the action of the radula

Preparation
None

Materials

Per student:
1. Sheet of glass, 25 cm × 25 cm
2. Specimen in container
3. × 10 stereomicroscope
4. Clamp
5. Paint brush
6. Container of flour
7. Container of water
8. Mixing dish for flour wash

Methods
Paint one side of the sheet of glass with a very dilute wash of flour and water. When dry, the glass should be faintly and evenly misted with flour. Put an active snail in the centre of the floured side of the sheet and wait until the animal is moving freely. Gently invert the glass sheet as in 18.1 and fix the glass in a clamp so that the undersurface of the snail can be examined through a stereomicroscope. The action of the radula can now be studied.

The snail will normally feed on the floured surface for some time, and this allows for easy observation of the action of the radula. Again, a number of elaborations are possible: the number of licks per minute, the area cleared per lick, the pattern of the feeding path, side-to-side 'scanning' motion of the head, the behaviour of the anterior tentacles, may be analysed in more detail.

Time
15-20 minutes

18.5. Location of food

Purpose
To provide the student with a number of lines of investigation into patterns of feeding in the snail.

Preparation
None

Materials

Per student:
1. Sheet of glass 25 cm × 25 cm
2. Specimen in container
3. ×5 lens
4. Container of dilute flour wash
5. Very fine brush
6. Pair of dividers
7. Ruler
8. mm graph paper

Methods
Paint a fine wavy line of dilute flour wash on the sheet of glass. With the sheet horizontal, place the snail so that it comes in contact with one end of the fine flour line. Record the subsequent feeding path, the scanning movement of the head, and the part played by the anterior tentacles.

This basic technique can be elaborated in a number of ways. The snail normally starts feeding when it comes in contact with the flour line and follows the line across the glass. Once this has been established, the obvious thing to do is to present the snail with different patterns of flour lines in such a way as to throw light on its feeding behaviour. For example, it is possible to present the snail with a 'Y' choice, where one arm of the Y is, perhaps, thicker than the other. Or, one arm of theY could be flavoured with lettuce extract. Another type of variation would be to find out if a snail treated a pattern of very closely painted parallel flour lines as a homogeneous surface and grazed across it at random. If so, how far apart must the lines be before the snail's feeding path follows the lines?

It is possible to find this distance and relate it to the area scanned by the snail as it feeds. The same sort of approach was devised by a school pupil who found that a snail would follow a line of flour dots on a glass plate. He reasoned that there must be a critical distance between the dots where the snail would no longer treat them as a line. He also predicted that this distance would be the same as the distance between the tips of the anterior tentacles and the mouth. He was correct in the first, but found that the distance was much larger than predicted, and eventually came to appreciate the function of the scanning behaviour of the head as the snail moves forward. Obviously, the open-ended nature of this rather convenient and economical type of investigation can be appreciated. But one other variation must be mentioned: when the glass sheet is clamped vertically rather than horizontally. Most terrestrial pulmonates show a strong negative geotaxis after immersion in water (or after rain in their natural surroundings) so that flour lines at different angles from the vertical can be tested. A snail will feed on a flour line which is vertical, but will show an increasing tendency to leave the flour lines the greater the angle becomes between the flour line and the vertical. However, as the snail dries out and the negative geotaxis weakens, the tendency to leave the flour lines diminishes with time. So here we have an interesting investigation possible which brings the student into contact with another behaviour pattern, a time problem, and an excellent situation for planning and discussing experimental designs.

Time
30 minutes upwards

18.6. Responses to different substrate textures and contours

Purpose
To introduce the student to simple methods of studying the responses of the snail to different substrate textures and contours.

Preparation
None

Materials

> Per student or small group of students:
> 1. Specimens in container
> 2. Pieces of gorse branches or similar spiny plants
> 3. Plasticine
> 4. Fine entomological pins
> 5. Plaster of Paris
> 6. Disposable Petri dishes
> 7. Clamp and stand
> 8. ×5, ×10 stereomicroscope or lens
> 9. mm graph paper

Methods
Release an active snail on a bed of gorse prickles and observe its locomotion. Formalize the problem by making up surfaces consisting of points at different densities. This is best done with fine pins set in plaster of Paris. Try surfaces made up of points at 1 mm square intervals, 2 mm, 3 mm and so on, until the snail no longer treats the points of the pins as a surface. Note how behaviour on such surfaces begins to change with the decrease in the number of points per square centimetre.

This approach is one of many possible on the topic of how snails cope with different substrates. The 'fakir's bed' of fine pins is best made by first inserting the pins into a holding device — a flat block of Plasticine is ideal (which has been marked with 1 mm squares).

A piece of damp graph paper laid on the Plasticine is all that is required. When all the rows of pins have been set into the Plasticine, point first at the required intervals, the Plasticine can be hardened in the refrigerator, which makes subsequent handling easier. The Plasticine is then clamped upside down with the heads of the pins in liquid plaster of Paris. After the plaster has set, and when the Plasticine has warmed up and softened, the Plasticine and graph paper can be removed, leaving an even surface of points firmly held in the plaster. At 1 mm intervals the apparent pattern of locomotion does not differ from normal movement on a solid surface. As the distances between the points increase, the student has the opportunity to study how the mucous trail is used, and how the snail begins to exploit the sides of the pins as well as the points.

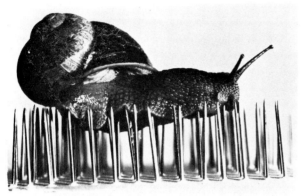

Figure 18.6.1 *Helix aspersa* moving over a plane surface made up of pinpoints at 2 mm intervals.

Other approaches to the substrate topic could be tackled along the lines of releasing snails in dishes containing different sizes of particles, from dusts and fine sands up through grits to larger gravel. The tests could be duplicated: one set dry, one set wet. At higher levels, the snails' responses to the biological and chemical differences in substrates (as well as their physical characteristics) can be tackled. And at all levels the study of the snails' reactions to different biological, chemical and physical barriers is another considerable source of the open-ended type of investigation (Townsend, 1973).

Time
1 hour upwards

REFERENCES

Townsend, C. R. (1973), 'The Food Finding Orientation Mechanism of *Biomphalaria glabrata*,' *Anim. Behav.*, 21, 544-548

19. Siamese Fighting Fish

INTRODUCTION

The Siamese fighting fish *Betta splendens* is a superb subject for laboratory behaviour work. It is attractive, easily kept in aquaria, and its display behaviour in particular is dramatic and easily observed.

The wild form of *B. splendens,* not often seen in aquaria, is distributed widely in Singapore, Malaysia and South East Asia in standing or slow-moving fresh water over a swampy bottom. The beautiful long-finned and highly coloured variants seen in the aquarists shop have been selected and bred from this wild stock. Sometimes called the Veil Tailed Fighter, these attractive variants are the animals most readily available to the experimenter. The fact that the fish have been carefully selected for colour and fin size must be kept in mind when considering the significance of the behaviour studied.

The ornamental variant shows the same pageantry of colours and poses as the wild type, but on a more dramatic scale. In general the wild *B. splendens* has a longer life-span than the cultivated form, which is at its best somewhere between four months and a year. Fish over a year old age quickly and become sluggish, therefore it is important to use young specimens in the laboratory.

These fish are most undemanding in the way of accommodation and food. They are easily kept in small containers, but they do require a constant high temperature if they are to be at their best. The aquaria should be between 28°C and 32°C. The tanks should be heavily planted, fully lighted from the top (sunlight is excellent) and covered with a sheet of glass. The glass cover is to ensure that the air above the water is also at approximately 30°C, because these fish are anabantids, a group characterized by the possession of a labyrinth or accessory breathing organ. Oxygen is taken in by gulping air at the surface, so it is important that this air is not chilled.

The sexes are easily distinguished. The females are less highly coloured and possess shorter fins, and a short but noticeable genital papilla.

Because of their aggressive natures, the males obviously should not be kept together. The fish are best kept in pairs or singly. The pairs will breed quite readily as long as they have a well-planted tank which is set up in a vibration-free spot. Since they are bubble nesters, it is important that the nest is not disrupted by movement of the surface water. The male looks after the spawn in the floating bubble nest until the young hatch out, and he will continue to look after the new young until he can no longer keep them in one place. It is usual to remove the female from the breeding tank once the eggs have been laid. The young should be separated and kept singly after four weeks.

EXPERIMENTAL WORK

19.1. Observation and description of displays

Purpose

Probably the most obvious feature of male Siamese fighting fish is the relatively enormous size and bright colour of the fins. In male-male encounters, these fins, the operculum and its associated branchiostegal membrane (gill cover) may all be raised. This is accompanied by an enhancement of the brilliance of the red or blue fin and body colour. As these displays are conspicuous, easy to elicit and composed of a number of fairly distinct components, this forms a suitable exercise to introduce the idea of the exaggerated ritualized nature of display behaviour. In addition it has a more general attribute which is to encourage accurate observation of behaviour.

Preparation

The aquarium tanks should be set up with a thermostatically controlled heating element and allowed to stabilize at about 30°C before any fish are placed in them. Standard aquarium tanks should be modified to allow two partitions to be placed across the middle of each tank, and these partitions should be in place before any fish are introduced. One male and one female fish should be placed in both ends of each tank, at least 24 hours before the practical class, to allow them to settle down. No food should be given to the fish for 24 hours before the class.

Materials

Per pair of students:
1. One aquarium tank about 45 cm × 25 cm × 25 cm separated across the middle by two removable partitions, one opaque and one transparent.
2. Two pairs of Siamese fighting fish, one pair in each end of the tank.
3. Three 250-ml glass beakers.
4. A stop watch or clock.

Per class:
1. A sufficient quantity of *Tubifex* to supply each pair of students with a 'pinch'.
2. A few small hand nets for removing or transferring fish.
3. Two or three thermostatically-controlled aquarium tanks ready to take fish which may be removed from their home tanks during the class.

Methods

It is worth while before doing anything else to ensure that everyone in the class knows the names of the various fins; this avoids any possible confusion later. The pectoral and pelvic fins are most likely to be confused, because the latter are almost as far forward as the former. The pelvic fins, which are elongated and dark, are also much more conspicuous than the pectoral fins, which are unpigmented and transparent (figure 19.1.1).

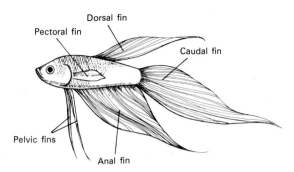

Figure 19.1.1 Diagrammatic representation of a male fighting fish, seen from the side with fins extended to show the location of the various fins.

The first observation on the display of the males should then be conducted with both the tank partitions in place, using only the display of the male to the female. The reason for suggesting this is that students unfamiliar with observing behaviour in general (and displays in particular) miss a great deal which may be quite clear to an experienced observer; therefore, to place two males together, which immediately causes very intense display and fighting, confuses rather than clarifies the students' ideas of display. The male alone with his female can be persuaded to display to her by putting some *Tubifex* in the tank; this, as well as inducing feeding, results in a general increase in activity and in display behaviour. The displays are, however, well spaced out and therefore more clearly recognized as separate elements. The names given here to describe the behavioural elements observed in this and other situations described are based on those of the monograph by Simpson (1968). The ones most likely to be seen by the male to the female are:

(a) *Fins spread:* Spreading of vertical fins
(b) *Gill cover erection:* Erection of gill covers and protrusion of the brachiostegal membrane (figure 19.1.2)
(c) *Bite:* Striking the other fish with the mouth and closing the jaws.

Of these displays it is the fin spreading, which is very conspicuous, that is most likely to be disregarded by students as a display. This is partly because it does not seem surprising to them that fish should raise their fins, and also because the fins may be spread when the male is 10 cm or more distant from the female. Males will, however, be observed swimming slowly towards

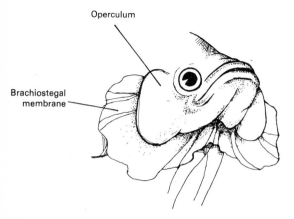

Figure 19.1.2 Head of male fighting fish with opercula raised and branchiostegal membranes extended.

or in front of females in a slightly head-down posture, stopping suddenly, and then fully extending all fins. After this has been observed 2 or 3 times, its ritualized nature will be appreciated.

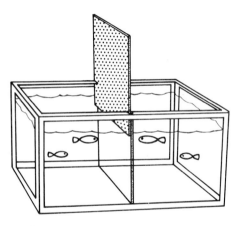

Figure 19.1.3 General view of tank containing 2 pairs of fish separated by a transparent partition, but with opaque partition raised.

The opaque partition across the middle of the tank should then be removed to allow the two males to see each other through the transparent partition (figure 19.1.3). This results in quite intense male-male interaction. *Fins spread* and *gill cover erection* will be observed as above. In addition students should observe:

    (*a*) *Tail beating:* the two fish swimming parallel, separated by the barrier and directing tail beats towards their opponent.

(*b*) *Colour change:* The red or blue colour of the males will be seen to darken and take on an additional brilliance.

At this stage it is important to encourage the observation of the finer detail of the displays. This may be done in a more general way at advanced levels or at more introductory levels by asking specific questions such as the following:

    (*a*) Is one male with fins spread in a particular body orientation relative to the other?
    (*b*) Is one male with gill covers erect in a particular body orientation relative to the other?
    (*c*) When pelvic fins are extended, is one extended further than the other; if so, which one?

These questions may be answered descriptively, or by the systematic scoring of the relevant situation observed several times. It will be found that *fins spread* tends to occur as a lateral display, and *gill covers erect* as a head-on display, one fish showing one while the other adopts the other (figure 19.1.4). Simpson (1968) noted that a pair of displaying males tend to alternate between one position and the other.

Figure 19.1.4 Two male fish engaged in mutual display. The fish on the left shows lateral display with fins raised, the one on the right shows head-on display with gill covers raised (after Simpson, 1968). Note that the fish in lateral display extend the pelvic fins unequally, the one nearest the opponent being raised more.

He suggested that the fish performing the lateral display was trying to regain the head-on orientation.

In the case of pelvic-fin-raising it will be seen that the fin nearer to the opponent will almost invariably be extended the further.

The transparent barrier across the centre of the tank should now be removed, so that the males may physically contact one another. In addition to those displays already seen, this allows the following additional ones to be observed:

(a) *Carouselling:* Both fish together slowly circling head to tail.

(b) *Mouth-to-mouth holding:* Both fish holding on to the other by the mouth, preventing it going to the surface for air.

Some pairs of fish will be found to be closely matched and will have eventually to be separated before they damage one another excessively.

In other pairs, one fish may quickly win a decisive victory, the defeated animal suddenly losing its bright colour, closing up its fins and retreating to a corner, after which the victor only displays to him occasionally.

It is worth looking briefly at female displays. It can be seen that all the fins on females are small and transparent, except for the pelvics which are pennant-like and pigmented, rather as in the males. These pelvic fins will be raised when females are approached by males or, when both barriers are removed, by the other female. This is probably a threat display as in males, but females do not seem to show any more intense interaction between each other than this.

A question which may well arise during the class is the possible function of these aggressive displays by male *Betta*. This is not an entirely satisfactory question to ask about a domesticated breed but, provided this is made clear, the situation may be used to provide discussion on the possible effect of domestication.

Female fish may be removed from the tank to test if males are displaying to protect their females. It will be found that male displays are apparently unaffected. To test whether males are displaying to defend a familiar home area or territory, one male may be introduced into a strange tank containing one unfamiliar male; both fish will again show vigorous displays.

Fish removed from the tank should be placed in marked beakers of water, and the beakers floated in the spare aquarium tanks to be brought back if needed.

These results suggest that, whereas wild *Betta splendens* may have fought to defend a home area or an area round a nest, the domesticated form is much more ready to display to other males at any time. Section 19.2 also demonstrates that males will display aggressively to a wide variety of stimuli.

Time

3 hours is required to carry out everything described; if additional time is available, some parts could be left for more detailed investigation during a further 1 or 2-hour period. It is desirable to allow additional time to discuss what a display is, how you would recognize it, and the possible effects of domestication, as well as the general observations and findings of the class.

19.2. Stimuli eliciting aggressive displays in male fighting fish

Purpose

A student, who has observed the intense display and often damaging fighting when two male *Betta splendens* meet, may be encouraged to investigate what features of the stimulus situation of a male displaying cause another male to start displaying. This may be tested using that classical tool of ethological investigation, models; in this case fish models.

Preparation

One male should be placed in a thermostatically controlled aquarium tank stabilized at about 30°C at least 24 hours before the investigation starts to allow it to settle down.

Materials

Per pair of students:
1. Three thermostatically-controlled aquarium tanks, 45 cm × 25 cm × 25 cm
2. Three male Siamese fighting fish; one in each tank.
3. Two stop watches, with separate stop-start and zero buttons.
4. Three pieces of card 20- cm square, one red, one blue and one black.
5. 40 cm of cotton thread
6. Two small paper clips

Methods

The display of *Betta* to models will be found to be not very intense, and quite variable between individual fish; for the latter reason 3 or more male fish should be tested with models. This exercise should also be carried out by students who have some familiarity with *Betta* displays such as would be gained by doing exercise 19.1. It should be begun by presenting to each test fish for 5 minutes a cut-out model of the lateral view of another fish with all fins raised (figure

19.2.1). The model should be of black card and presented outside the front of the tank on a piece of cotton thread and moved slowly up and down the tank. It is quite a good idea to mask the other three sides of the tank with paper.

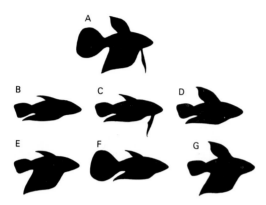

Figure 19.2.1 Silhouette of cut-out models of *Betta* males. Model A shows all fins raised, the rest show various combinations of raised and lowered fins (after Johnson and Johnson, 1973).

This initial observation allows students to observe the level of response before deciding what they are going to score. Johnson and Johnson (1973) who conducted a similar experiment scored the following:

(a) Latency of first approach.
(b) Total display time per test.
(c) Percentage of test time oriented to stimulus
(d) Rating for each gill cover extension on a 10-point scale.
(e) Rating for each fin spread on a 10-point scale.

These are useful measures for this exercise, with the following reservations. Firstly, unless a button-box recording device is used, it is not possible for a pair of students to record all these measures simultaneously if one of them is already engaged in moving the model up and down the side of the tank. Secondly, inexperienced observers will not sensibly be able to score (d) and (e) on better than a 5-point scale.

A range of models should be made in all three colours (such as shown on figure 19.2.1) to represent different possible combinations of fin raising shown by a male *Betta*. The models should then be presented, one at a time, to each fish according to a Latin square design. Each model should be presented for 2 minutes with at least a 10-minute pause between

each presentation. Models should be attached by paper clip to a length of thread; this allows a quick change of models.

The results of the experiment will be to show that there is no obvious enhancement of displays resulting from the extension of particular fins raised, but its statistical significance will probably be obscured by the overall variability of the response of this to the models. Colour of models will not be found to have any effect.

The lack of response to specific fin configurations and the slight enhancement of the response to the model presenting the largest surface area suggests that increase in surface area of the model may alone account for the increase in response. Fighting fish will respond simply to a circular cut-out moved up and down the tank, so the effect of surface area may be investigated using circular models of different sizes. The effect of the angularity of outline on the response may be investigated by comparing the response to a range of models of equal surface area but different angularity such as shown in figure 19.2.2. The effect of models of the same shape and area but different orientation may be investigated. We obtained some evidence in an undergraduate study that fish responded more to a black rectangle 4 cm × 2 cm when its long axis was horizontal than when it was vertical.

Figure 19.2.2 A range of cut-out model shapes of equal surface area to test response of males to angularity of the stimulus.

Many other questions may arise in an open-ended investigation, but it is worth mentioning the possible effect of recent experience on the readiness to display to a model. It was suggested at the beginning that at least 10 minutes should elapse between the presentation of models. What for example, would be the effect on the response to a standard model of a 15-second exposure to another male 2 minutes before the test as compared with a control? Would the response of the experimental fish be greater or less? This serves to draw attention to motivational factors which may be causing *short-term* changes in responsiveness.

The tendency of *Betta* males to display to a great

variety of shapes of models suggests that they would respond to a variety of species of fish other than their own. This can readily be shown to be the case. Johnson and Johnson (1973) investigated this in a fairly systematic way using two species of gourami *Colista*, the paradise fish *Macropodus* and a catfish *Corydoras* (figure 19.2.3). They found about as much display to all species as to another male *Betta*, except to the catfish which, partly because of its own lack of aggression and partly because of its inactivity and inconspicuousness, induced less display.

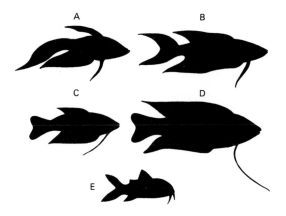

Figure 19.2.3 Relative sizes and shapes of stimulus fish. A – Fighting fish male *Betta*. B – Paradise fish *Macropodus*. C – Blue gourami *Colista*. D – Giant gourami *Colista*. E – Catfish *Corydoras* (after Johnson and Johnson, 1973).

Time
This is most suited to a study of project proportions. To perform the experiment with cut-out models of *Betta* alone will take 2 hours a day for about 3 days. This could be extended into 10 hours a week for 3 to 4 weeks if some of the other above suggestions were to be followed up as well.

19.3. Colour change in female fighting fish

Purpose
If a small group of female fish is kept in an aquarium tank with or without an accompanying male, it will be noticed that the females show a variety of colour patterns and that individuals can change quite rapidly from one pattern to another. This is an interesting example of colour change, but it also provides an oppor-

tunity for the contexts in which different patterns are shown to be investigated in order to arrive at some conclusion as to their cause and function.

Preparation
5 or 6 female fish, with or without a male, should be placed in a thermostatically controlled aquarium tank stabilized at about 30°C at least 24 hours before the investigation starts to allow them to settle down.

Methods
A lot of information may be gained by careful observation, even over a 5-10 minute period of time. This makes the exercise suitable for a simple demonstration of colour change and the context in which it occurs. This may then form the basis for a discussion on its significance.

The features which will be observed in a brief or extended period of observation are:

    (a) *Uniform dark colour:* Body colour dark red with iridescent highlights on the fins. This is shown by a female approaching another with its fins, and sometimes its gill covers, raised in the manner of a male displaying to another male. The response of the approached female is usually to fold up its fins and retreat. Dark colour seems therefore to indicate an aggressive state.

    (b) *Uniform pale colour:* Females of this body colour seldom raise their fins when approaching other individuals and retreat when threatened. The pale colour seems, therefore, to indicate a submissive state.

    (c) *Vertical bars:* Under certain circumstances, females will assume a colour pattern of alternating dark and light vertical bars. This is most clearly seen when the background colour of the fish is dark and the pattern appears as six or seven somewhat uneven, pale, vertical bars on the dark ground (figure 19.3.1). This pattern may be shown when a female is being bitten and chased by a male, or it may be induced by sharply striking the side of the tank. It is sometimes accompanied by 'freezing', where a fish stays rigid on the floor of the tank with pelvic fins fully extended for more than a minute. The pattern appears therefore to indicate a fearful state. It could possibly serve the function of making a fish less conspicuous to predators.

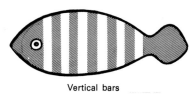

Vertical bars

Figure 19.3.1 Diagrammatic representation of female fighting fish showing dark body colour with pale vertical bars.

(*d*) *Horizontal bars:* This is a strong and obvious pattern when fully expressed and consists of two broad horizontal stripes, the lower one of which actually passes through the iris of the eye and ends at the tip of the snout. At the base of the tail there is a distinct dark spot (figure 19.3.2). Although this pattern is easy to recognize, the context in which it occurs is not so clear. It is shown by some females when they are making an apparently confident approach towards a male. It is shown by some females shortly after being chased by a male, changing from vertical to horizontal bars. The pattern may, however, occur when no males are present and be shown by females who are not apparently interacting with other individuals. This looks like an interesting problem for someone to investigate further.

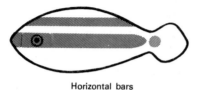

Horizontal bars

Figure 19.3.2 Diagrammatic representation of female fish showing two dark horizontal bars and dark spot at the base of the tail.

The above account may give the impression that the coloration of females is of four distinct types, but this is not the case. The patterns are not mutually exclusive, so that it is possible for a fish to be dark or pale while showing horizontal bars. It is also possible for patterns to show a gradation in their expression from slight to complete.

Fish do seem to be a class of animal that provides particularly remarkable examples of colour change and variety of possible colour patterns. Wickler (in Eibl-Eibesfeldt, 1970) observed colour changes in the cichlid fish *Hemichromis fasciatus* associated with territoriality, readiness to fight, spawning, care of young, hiding in the vegetation, hiding in the open. These colour changes involved the appearance of dark spots, light spots, horizontal or vertical dark bars, a dark eye stripe, and red spots or lines. This illustrates the kind of information that could be gathered for *Betta* by a careful observational study, even without recourse to experimental intervention.

Time
Some impression of the rate of colour change, different colour patterns and context of pattern could be demonstrated in 5-10 minutes; however, as an observational or experimental exercise, investigation could be of project proportions.

REFERENCES

Eibl-Eibesfeldt, I. (1970), *Ethology: The Biology of Behaviour* (Holt, Rinehart & Winston).

Johnson, R. W. and Johnson L. D. (1973), 'Intra- and Inter-specific Aggressive Behaviour in the Siamese Fighting Fish, *Betta splendens*,' *Anim. Behav.*, 21, 665-672.

Simpson, M. J. A. (1968), 'The Threat Display of the Siamese Fighting Fish, *Betta splendens*,' *Anim. Behav. Monogr.*, 1, 1-73.

20. Three-spined Sticklebacks

INTRODUCTION

The three-spined stickleback *Gasterosteus aculeatus* is a well-known pond, river and coastal fish and needs little introduction. It is familiar to most school children and is one of the few fish whose life history and breeding behaviour are known in outline by a surprising number of people outside academic circles due to the considerable volume of popular literature devoted to it. Largely through the efforts of Tinbergen, the three-spined stickleback has become a celebrated ethological species. His studies on male territoriality, nest building, courtship and care of the young are classics of their type (Tinbergen, 1951). Indeed, it is largely from these studies that our experimental work derives.

G. aculeatus has a very wide range, being found on the coasts and in the rivers of the arctic and temperate regions of the Northern Hemisphere, extending as far north as Greenland, Alaska and Kamchatka, and as far south as Japan, California, New Jersey and Spain. In northern regions it is essentially a marine fish; in the British Isles it is equally common in salt and fresh water; while in Spain and Italy it is almost entirely confined to fresh water.

It is easily maintained in small aquaria in which there is a bottom of loose sand and a number of plants growing, such as *Elodea canadensis*. The aquaria should be kept cool and away from direct side light. The sticklebacks are strictly carnivorous and require a diet of chopped white worms *(Enchytraeus)* and live *Daphnia* or other small crustacea.

Due to the time of year occupied by most courses a problem arises. Territoriality, nest building, courtship and similar studies must often be carried out at an inconvenient time of the academic year. However, sticklebacks do form loose agregates or schools when in non-breeding condition in the winter, and females in breeding condition form similar schools in early spring. These behaviour patterns may give scope for work with sticklebacks at a more suitable time.

EXPERIMENTAL WORK

20.1. Attraction of a stickleback towards a con-specific and schooling behaviour

Purpose

Outside the breeding season, the male and female three-spined sticklebacks are a mottled greenish colour, countershaded and of a generally inconspicuous appearance. During this time the fish are not territorial, but swim around in loosely aggregated groups of variable spacing and orientation between individuals. This apparent social attraction between individuals may be investigated in a simple laboratory situation.

Preparation

None

Materials

Per pair of students:
1. 4 three-spined sticklebacks in a home beaker or tank.

2. 1-4 ten-spined sticklebacks *Pungitius pungitius* (optional).
3. 1-4 minnows *Phoxinus phoxinus* (optional).
4. 1 opaque plastic trough about 50 cm x 25 cm x 15 cm or a small aquarium tank of similar dimensions.
5. 3 250-ml beakers
6. 1 small piece of water weed (any species)
7. A stop watch.
8. A small hand net.
9. A black Magic Marker.

Per class:
1. Some Plasticine.
2. Some fine stiff wire.

Methods

The experimental apparatus is set up as follows. The floor of the trough is divided across the middle by a black line down the inside of the tank. Each of the two halves should then be divided in half by lines drawn parallel to the first to give four segments in a row, each about 12 cm long by 25 cm wide. The trough is then filled to about 8 cm depth with water, and test stimuli placed one at each end of the tank in 250-ml beakers (figure 20.1.1). The stimuli are easly changed, and students can think of a variety of paired choices, given the equipment available. The test is conducted by placing a single three-spined stickleback into the tank at the mid-line, and then allowing it 2 minutes to settle down in the tank before scoring starts. The scoring system adopted may be devised by the students themselves, but a useful workable one is to score 2 for the left-hand side when the test fish is in the area containing the left-hand stimulus (L2) and 1 in the area (L1) immediately to the left of the mid-line. A corresponding scoring method is used for the right-hand side (figure 20.1.1). The fish may be watched for 10

minutes and its position scored every $\frac{1}{2}$ minute. The total left and right scores for the test then give a measure of the relative attraction of the two stimuli. All the fish should be tested, and appropriate controls included for possible end bias.

Using this apparatus the following sorts of preference should be observed (Table 20.1.1).

Table 20.1.1 The preference of three-spined sticklebacks for the pairs of test stimuli shown.

Stimulus 1	Preference	Stimulus 2
3 three-spined sticklebacks	>	nothing
1 three-spined stickleback	>	nothing
1 minnow	>	nothing
1 three-spined stickleback	≃	one minnow
1 live three-spined stickleback	>	one dead three-spined stickleback
1 dead three-spined stickleback	>	nothing
A piece of weed	>	nothing
1 three-spined stickleback	>	a piece of weed

These results show that a beaker containing 3, or even 1, three-spined sticklebacks is more attractive then an empty beaker. This could be due to some simple attribute of the stimulus fish, such as its shape or movement. A minnow is also preferred to nothing, and seems to be about as attractive a stimulus as a three-spined stickleback, suggesting the species body shape is not critical. The attractiveness of a live over a dead stimulus fish supported on a piece of wire suggests that movement is an important component of the attraction; however, even a dead fish seems to be slightly more attractive than nothing at all.

The attraction exerted by a piece of weed shows that three-spined sticklebacks may be attracted by stimuli quite unlike other fish in shape and movement, presumably because vegetation may offer some protection. One three-spined stickleback is, however, a more attractive stimulus than a piece of weed.

Approach behaviour
The approach behaviour of a test fish to the stimulus of a piece of weed may differ from its approach to another three-spined stickleback. It is interesting to note in addition that the three-spined stickleback is a bold species and will orient towards and may show cautious approach to a live pike *Esox lucius* in an aquarium situation. The approach behaviour of this species is therefore worth looking at in more detail.

Schooling
The attraction of one three-spined stickleback to another

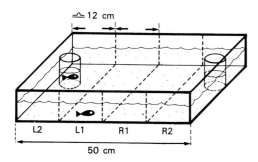

Figure 20.1.1 Apparatus to measure the attraction of a free-swimming fish to various stimuli placed in a beaker at one end of the tank (control beaker at the other end). Attraction is measured by the amount of time spent in L2 and L1 rather than R2 and R1.

suggests that schooling may take place in the wild. Limited observation on this may be made by releasing all four fish into the otherwise empty test tank. Observation of the fish will show that there is a rather loose grouping, but that individuals are inclined to move around rapidly and fairly independently. Group cohesion may in fact appear to deteriorate over the first 10 minutes' observation. These observations suggest that a tight schooling pattern, such as is seen in some marine pelagic species like the mackerel *Scomber scombrus,* is not characteristic of three-spined sticklebacks.

Comparative results for the ten-spined stickleback and the minnow

Tested with a choice of stimuli, the ten-spined stickleback will prove to be not very attracted to either conspecifics or to weed. The latter result is rather surprising, as the life of the ten-spined stickleback seems to be closely associated with aquatic vegetation which it uses as a refuge and nesting place, because it is so much less well protected from predators by its spines (figure 20.1.2) (Morris, 1970; Tinbergen, 1956).

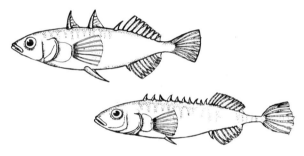

Figure 20.1.2 The three- and ten-spined sticklebacks, illustrating the difference in protection afforded by the long sturdy spines of the former and the relatively shorter and weaker spines of the latter.

The poor attraction to vegetation in the test situation may be due to the animal's fear, resulting in freezing and limited locomotion.

The minnow will prove to be strongly attracted to a beaker containing another minnow and, if four minnows are released in a tank, they show cautious movement in a closely packed formation or line ahead along the side. This schooling could serve as a further line of investigation, but requires more sophisticated apparatus beyond the scope of this book (Cullen, Shaw and Baldwin, 1965).

Time

In a $1\frac{1}{2}$-$2\frac{1}{2}$ hour class, each pair of students should be able to test systematically 2 or 3 pairs of stimuli. The various results of the class may then be brought together in a discussion.

20.2 Attraction of a stickleback towards a conspecific; more sensitive method for extended studies

Purpose

The purpose is the same as for 20.1, but in this case somewhat more sophisticated apparatus allows more sensitive measurement of the response of the test fish to a conspecific or other stimulus.

Preparation

A large round trough needs to be constructed before the class for each pair of students involved. It is suggested that this be made of heavy-duty polystyrene sheet about 3 mm thick. The trough should consist of a circular base about 60 cm in diameter with a wall 10 cm in height stuck onto the base with a suitable glue. The base should be marked out on the inside with concentric circles, the smallest of 5 cm radius and increasing by 5-cm steps. The floor should also be marked off in six segments of 60°. The five areas between the circles and the six segment areas should be labelled in a manner similar to that shown in figure 20.2.1 giving 30 separate areas.

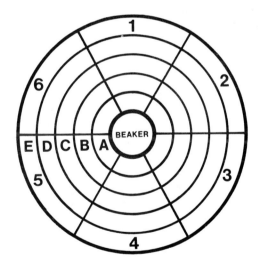

Figure 20.2.1 Plan view of an arena situation to measure the position and locomotion of a fish in response to various stimuli placed in a beaker which stands in the central ring of the arena.

Materials

Per pair of students:
1. 5 to 10 three-spined sticklebacks in a home aquarium tank.

2. 5 ten-spined sticklebacks.
3. 5 minnows.
4. 1 circular tank 60 cm diameter and 10 cm deep.
5. Three 250-ml beakers.
6. Some small pieces of water weed.
7. A stop watch.
8. A small hand net.

Per class:
1. Some Plasticine
2. Some fine stiff wire

Results

The attraction of a test fish to a conspecific or other stimulus is determined by placing a single stimulus in a beaker which is placed in the centre of the arena. The depth of water in the arena should be 5 or 6 cm, and into this should be released the test fish. The attraction of the fish to the stimulus may be determined by recording the position of the fish in the tank at intervals during a fixed test period, and comparing this with the behaviour of the fish to a different stimulus placed in the centre of the tank. The results will be essentially the same as those obtained in Section 20.1. The apparatus should, however, allow a more-sensitive measure of preference, even though there is not a simultaneous presentation of two test stimuli. This is because the test fish may swim freely around the test beaker without being disturbed by contact with the wall of a tank, and because the dividing of the floor area as shown in figure 20.2.1 gives much more information about the movement of the test fish than the rectangular tank in figure 20.1.1.

Time

A systematically conducted experiment demonstrating the attractive effect of the presence of a fish in the central beaker as compared with an empty beaker will take about 2 hours. This may be correspondingly extended by an open-ended study on the attractiveness of other kinds of stimuli placed in the beaker.

20.3 Reproductive behaviour of the three-spined stickleback

Purpose

The reproductive behaviour of the three-spined stickleback has become a classic example of certain important behavioural concepts. The courtship sequence from male zig-zag dance to egg fertilization exemplifies a stimulus-response chain culminating in a consummatory act. The zig-zag dance itself is said to represent a signal derived from intention movements of approach and withdrawal in a conflict situation (Tinbergen, 1952). The courtship by males of crude models of females, and their attack of models very unlike males except for a red patch, illustrates the idea of sign stimuli (Tinbergen, 1951). These theoretical points may be better remembered if students have actually seen the behaviour themselves. Apart from this, it is worth observing for its own sake as a remarkable insight into the private life of an animal; however, having given two good reasons for attempting such an exercise, it should be pointed out that it does have severe limitations. Firstly, the time of year when the investigation may be done is essentially confined to the normal reproductive season of the animal (May-June in Britain) because playing around with laboratory light and temperature is unreliable in bringing on reproductive condition. Secondly, experimental investigations using models are not to be recommended since, contrary to popular belief, a substantial proportion of male sticklebacks are quite unresponsive to models.

Preparation

Males and females entering reproductive condition come to the shallow edges of ponds, canals and rivers, and may be caught with a long-handled net.

Males in reproductive condition are easily recognized by the red colour of their bellies and iridescent blue eyes. The coloration of females will be similar to that of all sticklebacks in winter, but they will be distinctly plump about the belly (figure 20.3.1). A good quantity of aquatic vegetation should be collected at the same time as the fish.

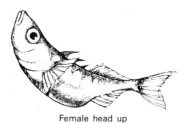

Female head up

Figure 20.3.1 A female three-spined stickleback in reproductive condition, as indicated by the swollen belly, adopts a head-up position in response to a male's approach (after Tinbergen, 1951).

In the laboratory, males should be set up either one or two to an aquarium tank 45 cm x 25 cm x 25 cm. The floor of the tank should be covered with clean, fine gravel or sand. A generous amount of aquatic vegetation should also be included, particularly if two males are to become established in such a small area; otherwise they will spend a great deal of time fighting, and one may be so aggressive as to prevent the other acquiring a territory. Females should be kept together in a separate tank.

Materials

Per demonstration:
1. One or two aquarium tanks 45 cm x 25 cm x 25 cm set up with male sticklebacks as described.
2. One aquarium tank with female sticklebacks.
3. A small hand net.
4. Some red and some grey Plasticine.
5. Some cotton thread.
6. A test-tube.
7. A small mirror about 10 cm square.

Methods

It is suggested that this exercise be treated as an extended demonstration which is set up in the laboratory over the three weeks or so of the reproductive cycle, so that students may follow the process by spending a few minutes in observation whenever they happen to be passing.

If a male is fully sexually mature it will start to build a nest on the same day it is introduced into its tank. Before starting to construct the nest, it will dig a shallow depression at the nest site by taking sand in its mouth and dropping it several centimetres away; it will then bring pieces of vegetation and press them into the excavated depression. The accumulating pieces of vegetation are stuck together with a kidney secretion which is applied to the nest in a movement called *glueing,* in which the body is dragged across the top of the nest. As the nest begins to develop into a distinct pad of vegetation, the male makes two nest openings by tunelling right through the middle of the nest material. After about two days the nest will be essentially completed. If two males are present in the tank, they will engage in vigorous fighting; if both manage to establish nest territories, fighting will then be confined to fairly well-defined territorial boundaries. Aggression takes the form of chasing, biting or a head-down threat posture (figure 20.3.2). The size of

Male head down

Figure 20.3.2 A territory-holding male three-spined stickleback responds to the approach of a rival male with a head-down threat posture.

tank suggested may prove too small for two fish to become established, in which case the nesting fish will periodically chase and bite the other, which will show no resistance.

If a male is in a tank on its own, then aggression towards an intruding male may be readily induced by placing a red-bellied male confined in a test-tube in the head-down position, into the tank. The resident male will also show marked aggressive behaviour towards its own reflection seen in a mirror. Tinbergen (1951) was able to provoke such behaviour from territorial males with models of various shapes, provided they had the stimulus characteristic of a red belly (figure 20.3.3). Such models can be made in a moment with red and grey Plasticine, but you should not rely on a response, since this is far from universal in male sticklebacks.

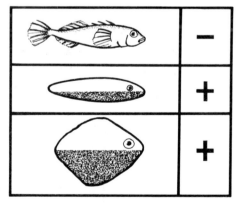

Figure 20.3.3 Three models used to test the stimulus eliciting aggressive response in a territory-holding male. The top model, which is very fish-like but without red belly, receives few or no attacks; the other two models, very unlike the sticklebacks in shape and general appearance but with red ventral areas, are attacked vigorously (after Tinbergen, 1951).

Tinbergen (1951) also showed that the swollen belly of the female was a sign stimulus to the male to release courtship behaviour; however, you may well find that any males you test are unresponsive to female models. The introduction of a ripe female will, however, immediately initiate vigorous activity in the male. At first he may be quite aggressive towards her, chasing and biting her; however, if she is receptive, she is not driven off but swims slowly near the nest in a characteristic head-up posture (figure 20.3.1). The male may then start to show a curious behaviour called *dorsal pricking,* in which he rises underneath her and pushes her upwards with his dorsal spines erect. After dorsal pricking, he may go to the nest and either *creep through* it or show *nest fanning,* the latter behaviour characterized by the fish facing the nest entrance and driving water through the nest with rapid beats of the pectoral fins while maintaining its position with beats of the tail. Wilz (1970) showed that these nest activities allowed the male to become less aggressive and suggested that dorsal pricking served the function of delaying the female's approach, thereby giving the male some time to reduce his aggression. You may be able to see that whereas, on introduction of the female, the male is very aggressive, later on, after he has performed some nest fanning and creeping through, he shows more courtship. The courtship behaviour takes the form of the *zig-zag dance,* a jerking swimming movement towards the female, followed by swimming towards the nest. This soon induces the female to follow the male to the nest, whereupon the male shows the nest entrance to the female with his snout; she responds by creeping into the nest. The male then rapidly vibrates his snout against her abdomen, causing her to release 50–100 eggs. The female emerges from the nest, and the male creeps through after her, fertilizing the eggs as he goes. He then drives off the female whose collapsed belly holds no more sexual interest for him.

During incubation of the eggs, which lasts about nine days, the male ventilates the eggs with periodic nest-fanning behaviour which becomes more frequent as hatching approaches. He also tends to cover the top of the nest with sand for reasons which are not clear. When the eggs hatch, the male at first retrieves any straying young in his mouth and spits them back into the nest. After a few days his retrieving declines and, about ten days after hatching, the young sticklebacks disperse.

The whole reproductive cycle is well shown in a film (Oxford Scientific Films) which may be hired from Film Libraries.

Time
Ten minutes a day, two or three times a week, over three weeks; all untimetabled except the courtship and egg-laying.

REFERENCES

Cullen, J. M., Shaw, Evelyn and Baldwin, H. A. (1965), 'Methods for Measuring the Three-dimensional Structure of Fish Schools,' *Anim. Behav.,* 13, 534–543.

Film: 'Behaviour of the Three-spined Stickleback,' Oxford Scientific Films, distributed by E. R. Skinner, Oxford.

Morris, D. (1970), *Patterns of Reproductive Behaviour,* London, Cape.

Tinbergen, N. (1951), *The Study of Instinct,* Clarendon Press, Oxford.

Tinbergen, N. (1952), 'The Curious Behaviour of the Stickleback,' *Scient. Am.,* 187, (6), 22-26.

Tinbergen, N. (1956), 'The Spines of Stickleback (*Gasterosteus* and *Pygosteus*) as means of Defence against Predators (*Perca* and *Esox*),' in Tinbergen, N. (1973), *The Animal in its World,* Vol. II, Allen & Unwin.

Wilz, K. J. (1970), 'Causal and Functional Analysis of Dorsal Pricking Nest Activity in the Courtship of the Three-spined Stickleback *Gasterosteus aculeatus*,' *Anim. Behav.* 18, 115-124.

21. Domestic Chicks

INTRODUCTION

The general biology and behaviour of the domestic hen *Gallus domesticus* is probably familiar to most people, even if they come from places where the hen is more commonly found in a broiler house than free-ranging. There are, however, one or two points worth emphasizing about the behaviour of the species, with particular reference to the chicks.

The probable ancestor of the domestic fowl is the Burmese Red Jungle Fowl whose general behaviour, including that of the chick, was described by Kruijt (1964). The Jungle Fowl is one of a large number of species of ground-nesting birds which have precocially developed chicks. On hatching the chicks are covered in fluffy feathers, have their eyes open and, within a couple of hours of hatching, are walking around and feeding themselves. This contrasts with

the chicks of tree-nesting species such as the passerines, where newly hatched chicks are blind, naked and helpless (figure 21.0.1).

Jungle Fowl chicks belonging to a particular brood remain in close association with one another and with their mother for at least a month. The role of the mother is initially to indicate the presence of food to the chicks and, for the whole period of maternal care, to warn chicks of approaching danger. The cock bird plays no part in the care of chicks. Essentially the same hen and chick behaviour has been observed in feral domestic fowl (McBride, Parer and Foenander, 1969).

Chicks are quite easy to obtain from local hatcheries. They are cheap if you get cockerel chicks. They are easy to keep, as they do not readily escape and don't mind being crowded. Newly hatched chicks can be kept for 48 hours without feeding, while they use up the remnants of the yolk sac. This initial period of survival without food is experimentally useful, because it allows chicks to be tested for, say, preference for pecking at different coloured objects before they have learned to associate food with a particular colour. Chicks should, however, always be provided with water and be kept warm (about 25°C). Best results will be obtained in a class only if the chicks are sufficiently warm. A final but important point in favour of chicks is that students and pupils like to handle them and find them attractive to work with.

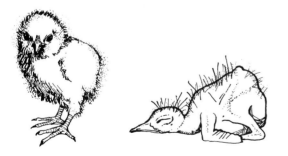

Figure 21.0.1 Left: One-day-old domestic chick showing eyes open, covering of body with down, and ability to stand. Right: One-day-old chick of passerine species showing its blind helpless state.

G

143

EXPERIMENTAL WORK

21.1 Approach of one chick to another

Purpose
To demonstrate a simple social response: that of the approach of one chick towards another.

Preparation
None

Materials

Per pair of students
1. 6 chicks, 1-3 days old
2. 2 cardboard or similar boxes of about 40 cm cube with a floor covering of sawdust to hold the chicks in
3. 2 60-watt bench lamps to keep the chicks warm
4. A stop watch or clock
5. 2 30-cm squares of stiff card
6. 50 g Plasticine

Per class
1. An assortment of Magic Markers of at least two colours.

Methods
All the chicks should first be marked somewhere on their feathers with an individually distinguishing mark using a thick felt 'marker'. One chick is then placed on the bench top or floor, and another placed about 25 cm away from it. After a short pause, and possibly some distress calling, one chick will approach the .other. This very simple and easily elicited response can be the starting point for a variety of open-ended lines of investigation. Having chicks running around on the benches and floor also provokes a lot of student interest and motivation, so the main problem is that of channelling the interest into the discipline of designing controlled experiments. The following questions are ones which may well be raised and which can be easily investigated.

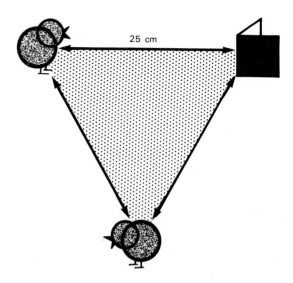

Figure 21.1.1 Triangular arrangement of chick, card, chick, spaced 25 cm apart to give each chick a choice of chick or card to approach.

(a) Does a chick prefer to approach another chick rather than an inanimate object of similar size?
   The inaminate object can be a 10-cm square of card stuck into a Plasticine base. A triangular arrangement of chick, card, chick, spaced about 25 cm apart, will then give each chick a choice between chick and card (figure 21.1.1).
   A chick will be found to greatly prefer approaching another chick, suggesting that the response is directed towards members of the species. Chicks will also approach their own reflection in a mirror. The species specificity of the response should not be too much emphasized, as a chick on its own will sometimes race towards the feet of someone shuffling slowly across the floor.

(b) Does the time taken for one individual to approach another decrease with repetition?
   Some individual chicks approach more readily than others, so that, if a pair is repeatedly separated, it is usually the same chick which approaches each time. The latency to approach can be measured on the stop clock and will be found to decrease rapidly over 5 or 6 trials. This modification of the behaviour can be shown not to be due to any familiarity with the approached chick, since changing that chick does not increase the latency again.

(c) Does a single chick tend to approach a pair of chicks?
   If two chicks are placed close together, they tend to stay close together. If a third chick is then placed 40-50 cm away, it is this latter chick that will approach the other two. This suggests that, to the single chick, the other two chicks are an attractive stimulus but, to one of the chicks of the pair, its partner is a more powerful stimulus than the more-distant solitary chick.

(d) Is the chick which distress-calls the most, the approacher or the approached?
   If pairs of chicks are placed on the floor about 50 cm apart, and for each approach a record is made of whether the most distress calling was done by the approacher or the approached, no clear correlation will be observed. It will also be noticed that there is no obvious effect of any kind of the distress calling of one chick on the behaviour of the other.

(e) Is the approach response entirely visually mediated?
   A pair of chicks can be placed on the floor about 50 cm apart, with a 30-cm square of card mounted on

a Plasticine base between them so that they are visually isolated. Both chicks will very likely show some distress-calling and begin to move around. No obvious response will be seen by one chick to the vocalization of the other; however, when they eventually move into position where they can see one another, approach follows rapidly.

This situation is such a flexible one that many other questions may be generated in a class allowed an open-ended investigation. Many of these questions may be investigated as simply as those described above.

Time

The approach response can be demonstrated in a couple of minutes. A worthwhile class investigation where each pair of students is required to set up and carry out a single controlled experiment could be carried out in 40 minutes. At more advanced levels (introductory undergraduate) there is enough information to warrant $1\frac{1}{2}$ to 2 hours. As in other open-ended class studies, it is rewarding to have a discussion afterwards, because different approaches will have been attempted to tackle the same question, and sometimes contradictory results obtained.

21.2 Distress calling

Purpose

As distress calling is a very easily and consistently evoked behaviour, it is very simple to set up experiments to investigate those stimuli that provoke it and those that inhibit it. It is not really feasible in most class situations to have mother hens present to see if their behaviour is influenced by chicks' distress calling; however, it is possible to investigate the influence of such calling on other chicks as shown in Section 21.1, Methods (*d*).

Preparation

None

Materials

Pair of students:
1. 6 chicks 1-10 days old in a home box 50 cm x 50 cm x 30 cm
2. 2 cardboard boxes about 40 cm cube for isolating single chicks
3. 2 60-watt bench lamps to keep chicks warm
4. A stop watch or clock
5. A mirror approximately 15 cm square
6. A thermometer.

Per class:
1. An assortment of Magic Markers of two or three colours.

Methods

All the chicks should first be marked somewhere on their feathers with an individually distinguishing mark with the Magic Marker. One of the chicks should then be placed in a test box and, after a minute to allow it to settle down, its cheep rate for the loud distress calls counted over a half-minute period. Using this simple technique the influence of the following variables on cheep rate can easily be investigated:

(*a*) Presence of lit or unlit bench lamp in the box.
(*b*) Presence of mirror in the box.
(*c*) Presence of another chick in the box.

It will be found that placing a lighted bench lamp in the box reduces the calling rate as compared with when it is not lit. This is almost certainly due to the temperature difference, which can be measured with the thermometer – not the difference in light intensity, although this cannot be easily demonstrated. The presence of the mirror decreases the frequency and intensity of the calling, and the presence of another chick may extinguish distress calling altogether.

Although chicks exhibit distress as measured by the calling when in the presence of other chicks, Section 21.1, Methods (*d*) showed that such vocalizations did not attract other chicks. Students may therefore infer that distress calling attracts the mother hen when a chick is in an unsatisfactory environment.

Time

Distress calling can be demonstrated in a couple of minutes, but can provide 30 minutes to 1 hour of open-ended investigation which should if possible be followed by a class discussion.

21.3 Comparison of the behaviour repertoire of one and nine-day-old chicks

Purpose

At its simplest level this is an exercise in observation; however, having observed closely the behaviour of chicks of two age groups, students should appreciate that, although many of the behaviour patterns are shared by the two groups, the older chicks show certain behaviour patterns not present in the repertoire of younger chicks. This raises more fundamental questions of how these new behaviour patterns appeared – by learning or by maturation of sensory, central nervous or muscular systems. These questions cannot be answered by this brief exercise, but it generates a situation where they may be discussed.

Preparation

All chicks should be food and water-deprived for 2-3 hours before the class.

Materials

Per pair of students:
1. 6 1-day-old chicks in a home box 50 cm x 50 cm x 30 cm the floor of which is covered with sawdust
2. 6 9-day-old chicks in a similar separate home box
3. 2 60-watt bench lamps
4. 2 small dishes of chick food
5. 2 small dishes of water.

Per class:
1. An assortment of Magic Markers of 2 or 3 colours.

Methods

At the start of the observation period on each of the age groups, the food and water should be placed on the box. This will have the effect of promoting many kinds of activity in addition to feeding and drinking. Students should watch each of the groups and take notes for at least ten minutes (and preferably longer) before attempting to draw up a list of behaviour patterns which represent the repertoire of behaviour patterns shown by the chicks. At a school level it is recommended that a check list of behaviour patterns is provided, and the number of times each one is seen scored for each of the two groups. At undergraduate or equivalent level, students can reasonably be asked to devise their own check list from scratch, so as to allow them to solve for themselves the problem of dividing up the behaviour record into meaningful and useful units.

Table 21.3.1, although not an exhaustive list, gives an idea of the kinds of behaviour that can be recognized at this level.

The discovery of water or food by the chicks gives rise to immediate feeding and drinking, accompanied by soft *pleasure twitters*. *Head shake* and *beak wipe* occur during feeding and drinking as beak cleaning

Table 21.3.1 The most easily recognized behaviour patterns shown by chicks, indicating the similarities and differences in behaviour between one-day and nine-day-old chicks.

	one day old	nine days old
(1) Walk	+	+
(2) Peck at ground	+	+
(3) Drink	+	+
(4) Beak wipe	+	+
(5) Head shake	+	+
(6) Wings preen	a little	+
(7) Both wings raised	absent	+
(8) Extend leg and wing on one side	a little	+
(9) Ground scratch	absent	+
(10) Pleasure twitter	+	+
(11) Distress call	+	+
(12) Alarm buzz	+	+
(13) Rest standing up	+	absent
(14) Rest sitting down	absent	+

movement, but can be precipitated by pushing a chick's head into water or sticking a small piece of adhesive tape on to the beak. *Distress calls* occur in this situation when a chick is a bit cold (see 21.2).

The *alarm buzz* is given when, as not infrequently happens, one chick is pecked on the face or feet by another. It can, however, be easily elicited by rapidly dipping a hand into the box of chicks and pulling one out.

Ground scratch is the most distinctive behaviour difference between the two groups. This is best elicited by scattering some food among the sawdust on the floor, although chicks will ground-scratch on a floor covered with food!

Preening and stretching movements, like *both wings raised* and *extending leg and wing on the same side,* are most clearly seen after feeding and drinking and before sleep.

Time

40-60 minutes; however, additional time may be necessary for a class discussion.

21.4 Pecking facilitation

Purpose

To demonstrate and investigate the social facilitation of the pecking response – social facilitation being the tendency of an animal to show a particular behaviour pattern in response to another animal doing the same thing, e.g. yawning in humans.

Preparation

The chicks should not be more than 2 days old and should have had no feeding experience before the class.

Materials

Pair of students:
1. 6 1-day-old chicks
2. 2 clean cardboard boxes 50 cm x 50 cm x 30 cm (one home box and one experimental)
3. 2 60-watt bench lamps to keep the chicks warm.

Methods

Recently hatched chicks with no feeding experience will peck at a variety of objects of small discrete appearance. This provides a situation where facilitation may be observed. If two chicks are placed in a box with a few small spots marked on the wall or floor, then the chicks will peck at them. The response of one chick pecking will attract the attention of the other chick to the same spot, at which it too will peck.

Approach and pecking can be experimentally elicited by tapping the floor of the box with a finger or pencil. This observation can be used as the starting point for an investigation of the stimuli facilitating pecking. If tapping *movements* are directed towards the floor without making the tapping noise, little or no response will be observed in the chicks; however, taps on the outside of the box, unseen by the chicks, facilitate pecking at a nearby spot on the wall or just at the wall itself near the origin of the sound. Similar results of the effect of a tapping sound were obtained by Tolman (1967).

Time

These effects can be demonstrated in about ten minutes and provide an experimental exercise lasting up to 40 minutes.

REFERENCES

Kruijt, J. P. (1964), 'Ontogeny of Social Behaviour in Burmese Red Jungle Fowl', *Behaviour*, Suppl. 12, 201 pp.

McBride, G., Parer, I. P., and Foenander, F. (1969), 'The Social Organisation and Behaviour of the Feral Domestic Fowl', *Anim. Behav. Monog.*, 2(3): 1-180.

Tolman, C. W. (1967), 'The Effects of Tapping Sounds on Feeding Behaviour of Domestic Chicks', *Anim. Behav.*, 15: 145-148.

22. Wood Pigeon Nests

INTRODUCTION

The nests of birds are a lasting record of some aspects of the behaviour peculiar to that species. Therefore the structure of a nest may be investigated in order to deduce the rules for nest building in the same way as suggested for caddis larval house building in Section 11.2. In some studies on the nest building of birds, for example Hinde (1958) on canaries, not only has the nest structure been described in some detail but also, by means of various experimental manipulations, the mechanism underlying nest building has been partially elucidated. This kind of investigation is not really feasible for the level of study dealt with in this book, firstly because canaries and other species do not readily build nests in laboratory situations and, secondly, because, at least in Britain, any direct interference with the bird itself, such as hormone injections, requires a Home Office licence.

The alternative to the study of nest building by captive birds is to look at nests built by birds in their natural habitat. In temperate regions, the location of the previous summer's birds' nests is suddenly revealed by leaf fall in the autumn, when the nests themselves are still in quite good condition. This is therefore a good time to collect the nests for detailed study in the laboratory.

The type of nest described here is that of the wood pigeon *Columba palumbus* which is a common British species. This builds a nest which with a little over-simplification may be described as a pile of twigs. The nests of many other bird species may be used for this study, as available, but the wood pigeon's is described here because its nest has certain features which make it particularly suitable:

(a) It is composed of comparatively few pieces.
(b) The pieces can easily be separated, weighed and measured.
(c) The nest pieces roughly represent the chronological sequence of their fitting from the bottom of the nest to the top.

The wood pigeon builds its nest above the ground in hedgerows or woodland, in a great variety of trees, evergreen, deciduous and coniferous. The nest may be from no more than 2 metres to more than 20 metres above the ground, quite hard to see during summer, but often clearly visible after leaf fall. Nesting is not colonial, so nests are somewhat scattered, but not uncommon when they occur at all. They have been recorded up to 30 nests/acre (Murton, 1965). Both members of the pair are involved in the nest building, often with one bird collecting the twigs, while the other incorporates them into the nest. Twigs are usually brought to the nest more than one at a time. Nests may be built on the foundations of a previous year's nest, in which case the earlier nest is tidied and rearranged somewhat before the new building starts. Nests which are built upon previous ones in this way usually result in a somewhat larger structure overall (Murton, 1965).

EXPERIMENTAL WORK

22.1. Investigation of nest structure

Purpose
The collection and study of completed nests, of

148

course, rule out any possibility of experimental intervention, however they do allow:

(a) Detailed observation and description of nest structure.
(b) Examination of the variability of nests taken from the same type of habitat.
(c) Comparison of the differences between nests taken from different habitats (e.g. coniferous and deciduous woodland).

This allows general statements to be made about the preferred number, length and weight of twigs in a nest, the individual variability in building standard imposed by availability of building materials. The exercise requires systematic weighing and measuring, and the tabulation of simple data; however, it is not very suitable for large classes, as nests cannot be collected in large numbers.

Preparation

The first requirement is to collect the nests, but this may effectively be used as part of the practical exercise by taking students equipped with a short ladder into woodlands and hedgerows in search of nests. This allows them to see the natural location and variation in location of nests.

Materials

Per two or three students:
1. A balance to weigh to the nearest 0.1 g
2. A 30-cm rule
3. 2 plastic trays about 50 cm × 30 cm.

Methods

The method of investigation of the nest assumes that the completed nest represents approximately the chronological sequence of pieces incorporated into the nest, starting at the bottom and working up to the top of the nest; this is probably an oversimplification, but does not affect the value of the exercise. In practice it is easier to study the nest in reverse chronological order, dismantling the nest from the top to the bottom.

The topmost twig of the nest is first removed, measured and weighed, and possibly scored as to whether it is branched or unbranched. The next twig is then removed and scored, and so on to the bottom of the nest. Accurate weighing is, of course, easy and may sensibly be done to the nearest 0.1 g, but estimation of length is much more inaccurate, so it is

suggested that twig length is estimated only to the nearest 5 cm, disregarding all except the longest branch and assuming that it was straight.

It will be found on dismantling the nest that, although it looks as if it has been constructed by piling twigs one on top of the other, the twigs have in fact been inserted into the nest, so that one end of the twig has been tucked under ones which had already been incorporated. This makes it more difficult to decide on the actual order of twig incorporation, but removing the twig which seems most free from entanglement, and taking twigs from the centre and working out to the edge for each 'layer' of nest, probably gives a reasonable approximation of the chronological sequence. Since not too much weight should be placed on the exact twig order, it is better to remove twigs in sequences of ten, lay them out on a tray, measure them individually, and then weigh them all as a group, and calculate the mean twig mass and length for those ten.

The findings discussed here are for a single nest taken from a larch plantation.

(a) *Building materials*
The nests are made mainly of twigs alone or incorporating some flexible woody stems or roots of trees and shrubs. The species of plant from which these are taken are quite varied. This nest was composed of twigs of the European larch *Larix decidua,* and stems and roots of bilberry *Vaccinium myrtillus.* Other nests taken from the same coniferous plantation were made of twigs of larch and Norway spruce *Picea abies.* Nests from deciduous woodlands are again predominantly twigs but of locally available species: oak *Quercus robur;* beech *Fagus sylvatica,* or hazel *Corylus avellana.*

(b) *Position of the pieces*
Dismantling the nest reveals that individual twigs are not laid directly on top of one another, but tucked underneath already attached twigs. As even the straightest twigs have several small side projections (in the larch these are prominent and numerous (figure 22.1.1)) then the effect of the tucking in of twigs is to produce a surprisingly rigid structure. It will also be seen that most twigs are laid tangentially and not radially to the centre of the nest (figure 22.1.2).

(c) *Overall mass of twigs*
Plotting the overall mass distribution of the twigs (figure 22.1.3) shows that practically all twigs weighed less than 2.0 g and that the modal mass was also the lightest, i.e. 0.1 g. Mean twig mass was 0.5 g.

(d) *Overall length of twigs*
The distribution of twig lengths was very different from that for the masses (figure 22.1.4) with a single peaked distribution with a mode of 20 cm and a mean twig length of 20.5 cm. The fall off either side of the peak is

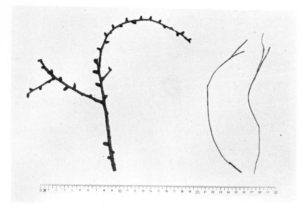

Figure 22.1.1 (Left) Twig of Larch taken from the structural part of the nest, showing short side shoots which help to hold the nest together. (Right) Long but thin and flexible plant stems illustrating the nature of the lining materials of a pigeon nest (scale in centimetres).

Figure 22.1.2 Pigeon nest from below to show tangential arrangement of twigs in the structure.

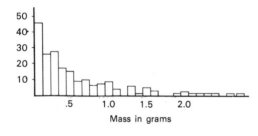

Figure 22.1.3 Overall distribution of twig mass in a single nest.

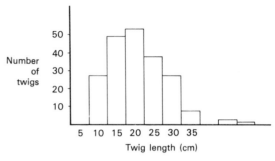

Figure 22.1.4 Overall distribution of twig length in a single nest.

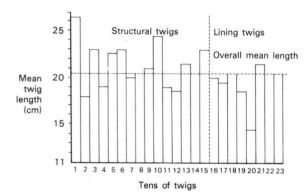

Figure 22.1.5 Mean twig length per ten twigs from bottom to top of nest. The horizontal broken line represents overall mean twig length for the nest, and the vertical broken line indicates the point of change-over from structural to lining twigs.

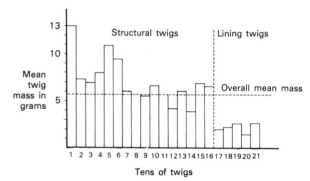

Figure 22.1.6 Mean twig mass per ten twigs from bottom to top of nest. The horizontal broken line represents overall mean twig mass for the nest, and the vertical broken line indicates the point of change-over from structural to lining twigs.

quite rapid; no twigs were found in the 5-cm class and hardly any above the 35-cm class.

(*e*) *Changes in twig length during building*
The histogram of mean twig length per ten twigs from the bottom of the nest to the top (figure 22.1.5) shows that, although there is possibly a slight decline in twig length in the later stages of building, the overall pattern is of a fairly steady selection of lengths about the mean.

(*f*) *Changes in twig mass during building*
The histogram of mean twig mass per ten twigs from bottom of the nest to the top (figure 22.1.6) shows a rather different picture to the change in twig length. The mass of twigs fitted drops suddenly below the overall mean for the last 50. This marks the change

from *structural* to *lining* materials; in this case a change from sturdy rigid larch twigs to thin flexible bilberry stems. The absence of a fall in length of twig corresponding to the change in materials shows that the lightness of the lining materials is caused largely by a decrease in *width* of the pieces, not their length (figure 22.1.1). This change in materials in the later stages of building is a feature of wood pigeon nesting. The tangential arrangement of the structural elements does in fact cause a slight depression in the centre of the nest, which is then filled in with the lining elements. Similar changes in building materials for the lining of the nest cup are very widespread in birds, and it was in fact the change in building material selection of canaries from grass to feathers that was one of the main features of the studies of Hinde (1958) and Hinde and Steel (1962).

It is useful in this kind of study to be able to pool the results obtained from more than one nest in order, for example, to compare nests from deciduous with those from coniferous woods.

It is, of course, very unlikely that these nests will have the same number of twigs, especially as this appears to be quite variable (between about 150 to 250), but the nests need to be aligned according to some common behavioural marker in time. This could conveniently be any of the following:

(*a*) The fitting of the first twig.
(*b*) The fitting of the last twig.
(*c*) The fitting of the middle twig.
(*d*) The fitting of the last structural twig.

All these represent events which indicate some common state shared by different builders, but equally there must have been individual differences, or all pairs of birds would have fitted the same number of twigs. If the nests are aligned by any of criteria (*a*), (*b*) or (*c*), the occurrence of the change from structural to lining material will be to a greater or lesser extent obscured. This change is, however, a marked and consistent one, and may best be emphasized by adopting criterion (*d*) for aligning the nests.

Time

This exercise is rather better suited to a project type of study. It will take about $4\frac{1}{2}$ hours for two students to dismantle a nest of 200 twigs, then weigh, measure and make general notes about them (this is assuming that they have a single-pan direct-reading balance). If a two-pan balance is being used, the exercise will take rather longer. Another 2-3 hours will be required to work out mean masses and lengths, and to tabulate results. The project time will therefore be correspondingly increased for each additional nest investigated.

REFERENCES

Hinde, R. A. (1958), 'The Nest-building Behaviour of Domesticated Canaries,' *Proc. zool. Soc. Lond.*, 131, 1-48.
Hinde, R. A. and Steel, E. A. (1962), 'Selection of Nest Material by Female Canaries,' *Anim. Behav.*, 10, 67-75.
Murton, R. K. (1965), *The Wood Pigeon*, Collins (New Naturalist).

23. Shrews

INTRODUCTION

Traditionally shrews have been regarded as difficult animals to trap and keep alive; they have thus been neglected as suitable experimental subjects. However, thanks to pioneering work of researchers such as Crowcroft, and the development of the Longworth Small Mammal Trap at Oxford, it is now clear that shrews can be caught easily and, with a suitable feeding regime, can be kept for many months in the laboratory. They make fascinating and most rewarding subjects for behaviour studies.

Shrews of the genus *Sorex* are widely distributed throughout the land masses of the Northern Hemisphere. The taxonomy of the genus *Sorex* has become complicated. In the common shrew *Sorex araneus* there exists remarkable chromosomal polymorphism. Two basic phenotypically indistinguishable forms, A and B, are now recognized. The common shrew of Britain is form B (see van den Brink, 1973). This form is found in Western Europe from France to Finland. In the British Isles the common shrew is numerous in a wide variety of habitats, but it is absent from Shetland, Ireland, and the outer Hebrides. A more widely distributed but less numerous species in the British Isles is the smaller pygmy shrew *Sorex minutus*. This tiny and charming animal is less suitable than the common shrew for the type of experimental work described here, but when available could provide an interesting comparison.

The identification of shrews presents little difficulty. G. B. Corbet's *The Identification of British Mammals* (1964) or the *Field Guide to the Mammals of Britain and Europe* by F. H. Van den Brink (1973) are useful.

The Common Shrew is easily trapped in the Longworth Small Mammal Trap or similar live trap. Early morning trapping usually produces the most shrews. The traps need not be baited, but do require careful siting. The trapper must examine the ground carefully and place the trap across a likely shrew run. With experience, the trapper can quickly assess the possibilities of an area and somehow get the feeling for 'shrewy' ground. With unbaited traps it is most important that the trapper visits the traps at intervals of no more than two hours, and should carry a supply of food such as mealworms, so that the captured shrews may be fed. Shrews deteriorate rapidly after two or three hours without food, and frequently die before they reach their intended home tank in the laboratory. If the traps have to be left in position for more than a couple of hours, a supply of chopped earthworm and some oats should be left in the traps. This will ensure that the shrews do not die from lack of food or moisture during their stay in the trap box.

A suitable home for the common shrew is easily set up in an aquarium tank. A layer of moist garden soil about 10 cm deep and a couple of pieces of moss or moss-covered bark, perhaps a little leaf litter and a small quantity of dry grass or hay, is all that is required. The tank need not be covered if it is more than 35 cm deep; shrews are expert climbers but poor jumpers. If necessary the top of the tank can be covered with a lid of fine mesh wire. Avoid a glass or other cover which will restrict air circulation. In general give each shrew 30 cm × 30 cm × 30 cm living space; two shrews can be kept in a standard 60-cm aquarium quite happily if there is plenty of litter, moss and grass cover. Feeding must be carefully planned so that the shrews will get a little often, rather than a large volume all at once. The common shrew will consume rather less than its bodyweight of food in 24 hours. If this is given all at once, the shrew most likely will gorge itself on it within the first few hours, and actually die of starvation before the next feed the following day. Thus the plan must be to ration out the food as carefully as circumstances permit. Certainly three feeds a day should be regarded as the minimum frequency, and it is often a good plan to scatter some excess oats into the tank if they cannot be fed sufficiently late at night. Although shrews are largely carnivorous, and must be supplied with live foods such as earthworms, mealworms, beetles, tadpoles, bluebottles, maggots, as available, they also will take certain brands of dog foods mixed with an equal quantity of oatmeal. This provides a readily available staple diet. But care must be taken to ensure that it is completely fresh. An all-meat diet is unsuitable and, if live foods only are supplied, they must be supplemented with edible seed and oatmeal. Water, of course, must also be constantly available, and this is best accomplished by standard drip bottle.

EXPERIMENTAL WORK

23.1. Burrowing behaviour

Purpose

Common shrews show three more or less distinct forms of burrowing behaviour. These can be labelled (a) *ploughing*, (b) *tunnelling and* (c) *surface runway making*. All three can easily be studied in the laboratory with a minimum of preparation, and results obtained within an hour or, at most, overnight. The shrews normally show some form of burrowing behaviour within minutes of being introduced into the experimental tank; thus this makes a good reliable piece of work for the classroom.

Preparation
None

Materials

Per student or small group of students:
1. 1 small aquarium or observation tank

2. 1 bucket of moist sieved garden soil
3. A small bunch of dried grass or hay
4. Container with half a dozen mealworms
5. 1 common shrew in small escape-proof container.

Methods
(a) Ploughing
The procedure is simply to put a layer of moist soil 8-10 cm deep in the glass tank and level it out carefully so that the surface of the soil is smooth and even, but not pressed down too firmly. The shrew is then released in the tank and careful observations made of its behaviour.

Ploughing is a rather mole-like form of surface burrowing, where the shrew ploughs through the surface or just below the surface, so that the soil is raised in a ridge which subsides as the animal burrows forward. Frequently the shrew's snout will break up through the surface and then disappear again. This type of burrowing is most frequently seen in shrews exploring new ground, so the odds are very much in favour of seeing this ploughing behaviour before true tunnel construction begins.

Crowcroft (1957) showed that this behaviour could be triggered by sprinkling a settled home tank with water. The shrew would respond immediately to such 'rain' and plough vigorously. The food-finding advantages of such behaviour are obvious, but well-fed specimens also show ploughing, and the erratic violence with which it is carried out makes one think of play rather than food finding.

(b) Tunnelling
The procedure here is virtually the same as above. However, if the soil is pressed down more firmly (not solid) the shrew may well start deeper tunnelling sooner. Most frequently the exploring shrew arrives at a corner of the tank and starts to dig with a furious energy. It pokes its nose into the soil, digs with its paws like a miniature dog, its legs working at lightning speed, and ends up practically standing on its head as it burrows straight down. Soil showers out between and past its hind legs and, if the soil is not too firmly packed, the shrew is out of sight in a matter of seconds. Once tunnel construction is under way, the tank should be left undisturbed for a few hours, preferably overnight, so that a pattern of tunnels is completed. It is then a comparatively simple matter to remove the shrew (with a live-trap or old coffee jar)

and carefully pour a thin mixture of plaster of Paris and water into the tunnel system. The tank should be rocked slightly to dislodge air pockets and ensure complete filling of the tunnels. After a few hours (ideally, the longer left the better) the soil can be scraped away and a permanent cast of the underground burrow system obtained.

This technique for recording tunnel patterns makes it easy to develop an investigation of tunnel burrowing along project lines. To facilitate the removal of soil and tunnel cast, prior to scraping away soil and the examination and preservation of the cast, it has been found easier to use shallow plastic tanks, or deep trays with sides no higher than about 15 cm. A fine wire cover is then needed to prevent the shrew escaping, and two plywood boards have been found to be useful in handling the soil and plaster mass. When the shrew has been safely removed, the first board is placed over the soil and the tank carefully turned upside down. When the plastic tank has been lifted off the soil, the second board is used to turn the soil and plaster tunnels back the right way up again.

Investigations can be carried out along a number of lines such as:

(a) Does each shrew produce its own characteristic tunnel pattern?
(b) Does the tunnel pattern vary with the depth of soil or shape of the tank?
(c) What happens when two shrews are tunnelling in the same tank?

And there are obviously a considerable number of other questions and approaches possible.

Shrews do not seem to have a consistent pattern of tunnelling. If one shrew is allowed to complete a number of tunnel systems and the casts compared, it is difficult to find a fixed pattern apart from the fact that corners are typical entrances and exits, and that the outer burrows may consistently follow the side of the tank. An overnight tunnel system is no more or less complex than a system produced over a number of days. Crowcroft (1957) showed that when nesting materials were also present, the nest was built towards the centre of the tank, and that it was common for burrows to radiate out from a bolt-hole below the nest. The majority of our shrews built their nests at the side or in a corner of the tank, but always had bolt-holes beneath them.

If the position of the surface exits/entrances are mapped over a day or two, it is clear that the shrew is constantly modifying and changing the pattern of tunnels. A comparison of these surface maps with the final cast gives some idea of the amount of change in pattern produced over two or three days.

If we wish to investigate the method of digging and how the shrew copes with different substrates, obstacles, water in the substrate and so on, the following simple apparatus is useful. It is essentially two sheets of glass kept 3 cm (or any desired distance) apart by wooden spacers. The gap between the sheets is filled with the substrate required, and a wire mesh or tight-fitting Perspex or plastic entrance guide constructed so that a shrew in a small tank has access to the soil in the glass sandwich. This glass observation sandwich can be used laid flat on the bench, or vertically, or at varying degrees of slope as required. Therefore some sort of efficient clamping system is essential. Some students will also prefer a large reading glass for close observation of the burrowing shrew.

With this observation system it is possible to watch a digging shrew at close quarters when it comes up against water, a root system, or pebbles. A shrew normally picks up an obstructing pebble and backs out of the burrow to get rid of it. It is interesting to note that the shrew carries the obstruction in its mouth, and that it may wander about on the surface for some time before deciding what to do with such an object. This behaviour, again, is open to further investigation.

(c) *Surface runways*

The idea here is to restrict the shrew to a mass of typical loose surface materials such as moss, grass clumps and the like. The tank therefore should have only a thin sprinkling of moist soil as the bottom layer, and should contain a loose clump of material such as hay or dried grass and interwoven moss. The clump should be 15-20 cm in depth. A shrew is introduced to the tank and observed for the first ten minutes or so, and then again after several hours.

The initial exploratory behaviour of the shrew is very different to the behaviour observed later. At first the animal almost nervously tests and explores the hay. However, after two or three hours the shrew, when disturbed, flits like lightning through a system of loose channels, tunnels or runways it has constructed through the hay. Its behaviour is certain and confident. It seems to know exactly where it is going without having to check. This is borne out by an experiment such as simply putting an obstacle across one of the runways. Nearly always the shrew will collide head on with such an obstruction. It then has to investigate and tunnel round the obstruction. This new path is preferred for a time after the obstruction is removed.

Time
30 minutes upwards.

23.2. Feeding

Purpose
Although classified as a member of the Insectivora, it is misleading to think of shrews as insectivorous. They are carnivorous; any small invertebrate or vertebrate is potential prey. Beetles, spiders, earthworms, young frogs and fresh carcasses of field mice or voles are tackled with equal vigour. Shrews show all the determined ferocity of a terrier worrying a rat. In fact their feeding behaviour in general is dog-like. This is in direct contrast to the quieter nibbling, gnawing and cropping of the herbivorous rodents. This exercise is simply to observe shrew feeding behaviour with a variety of foods, so that a number of facets of such behaviour can be studied, a number of problems raised, and further possibilities of experimentation suggested.

Preparation
The shrews should not be fed for about an hour before the class.

Materials

> Per small group of students:
> 1. 1 aquarium tank set up as a shrew house with 1 resident shrew; for clear viewing reduce the surface materials, leaflitter, etc., to a minimum
> 2. Containers with various food items
> 3. 1 pair of forceps

Methods
The basic exercise is to feed the shrew with different sizes and types of food, and build up a picture of its feeding methods. Once this has been done, further feeding tests can be carried out to try to shed light on

problems thrown up during the initial observations.

A very hungry shrew will eat its newly caught prey more or less where it catches it. A beetle, for example, will be found by apparently stumbling over it more by accident than hunting technique; it will be bitten, worried and, if palatable, eaten. A less hungry shrew will take its prey home to the nest. Or, if the shrew is becoming satiated, the food may be buried or simply ignored.

It is difficult to see precisely how the shrew attacks and dispatches its prey because of the small size of the animal and the speed with which it does it. It was found that shrews could be persuaded to feed from a small Perspex observation chamber with removable windows made from microscope slides. The chamber was attached to the side of a home tank, and a low-power stereomicroscope and the necessary lighting arranged so that the feeding shrew could be observed at close quarters, from the side, from below, or from whatever angle might be required. It was also found necessary to anchor the beetles or other food item to the end of the chamber, otherwise the shrew would quickly snatch the food from the chamber and disappear with it at lightning speed. Cine or video-tape recording techniques are really required here to slow down the various components of the attack and dispatch behaviour. However, with an anchored centipede, or other longish prey species, one could observe with the unaided eye the method of quickly sniffing or touching all along the prey, and then quickly and accurately placing dissabling bites along the animal. With some beetles one bite in the head region was all that was required. With longer species, two or three evenly-spaced bites were noted. Then began the struggle to remove the prey to the nest. Some of our students contrived two-headed beetles to see how the killing pattern was altered; others studied the opening and feeding on vole carcasses, and so on. There are obviously a number of possible developments here.

Other feeding patterns will become apparent in the aquarium tank itself. Students will quickly notice that a shrew may drop its food and will continue on to the nest before returning to pick the food up again. It looks as if the shrew is making sure that the way is clear, or just reminding itself of the way home. An interesting comparison can be made between the neat stockpiling of wheat grains at the nest, and the laborious haulage of a sizable vole carcass to the vicinity of the nest. How a shrew deals with a ball of whiteworm *Enchytraeus* or *Tubifex*, which disintegrates on the way to the nest, also provides an illuminating problem. Again numerous experimental approaches are possible.

In one study of food location and memory of productive pathways, one shrew became accustomed to crossing a mossy tank, climbing a wire tube, crossing a wire bridge, descending a wiremesh tube into another tank, crossing a stone barrier, swimming across a small pond (180 mm across) and finally climbing up into a wire (Weldmesh) maze which it had to remember before it got its mealworms. And the shrew had to carry its laboriously earned prey all the way back to its nest (Jones, R. K. H., unpublished demonstration).

Time
30 minutes upwards

23.3. Nest building behaviour

Purpose
Shrews show a regular pattern of nest building and, since they readily build nests in captivity, this behaviour provides a useful observational and interpretive piece of work for students at a wide variety of levels. The initiation of nest building is unfortunately difficult to time for a class, but as a short project extending over a week or so it is particularly good. Even a fairly superficial analysis should bring out some of the essential features of this interesting behaviour pattern.

Materials

Per small group of students:
1. 1 tank with resident shrew. The tank should be comparatively free of potential nesting materials
2. Containers of grass, feathers, shredded paper, pieces of wool, chopped sacking.

Methods
The procedure is simply to present the shrews with one or more types of nesting material in varying amounts, and to record the components of nest-building behaviour as they occur and/or analyse the nest as it is being constructed and nests that have been built.

The typical shrew nest starts as a hollow platform which, if the supply of materials allows, becomes a well-made hollow ball. The building pattern is stereotyped.

(a) Pulls a ring of nesting material around itself.
(b) Turns round and round while holding pieces of nesting material in its mouth. This produces a loose wall of material round the animal. The nest builder slowly becomes embedded in a roughly spherical mass.
(c) Consolidates the nest by pulling the materials evenly in towards itself with its mouth. This eventually produces a tighter and neater sphere. The snout of the shrew continuously pokes through the sphere as it rounds it off.
(d) Modifies nest by opening entrances, repairing, changing runs until finally satisfied.

The nest is typically linked to the established tunnels by a bolt-hole. If one is lucky enough to have a pregnant shrew, the nest-building behaviour is considerably more intense.

The student will notice that the building is done with the snout and not the feet as with small rodents such as the house mouse. The nest building of shrews and other small mammals makes a useful comparative study. The topic has also been approached with success from another point of view. The delicate crocheting action of the long snout of the shrew is obvious during nest construction; the determined drilling action of the shrew snout during tunnel construction is also notable. A project concerned with the structure and functions of the shrew snout is therefore a possibility.

Time
Open-ended to about 10 days on average.

23.4. Fighting behaviour

Purpose
Patterns of fighting behaviour are relatively easy to produce for observation and interpretation by students. The technique described here will nearly always produce some part of the fighting pattern, and occasionally a considerable display involving most of the possible components known.

Materials

Per small group of students:
1. Long narrow Perspex or glass observation tank,

400-500 cm long, 250 cm high and only 100 cm wide.
2. 2 shrews in separate containers
3. Moist soil for base of observation tank
4. Sheet of Perspex 250 cm × 99 cm for dividing tank
5. Tape recorder for recording the shrew cries.

Methods

With 2-3 cm soil in the bottom of the observation tank and the divider in position, a shrew should be introduced into each half of the tank. Give the animals five minutes or so to settle down, and arrange the microphone of the tape recorder so that it is well down inside the tank. Finally lift out the tank divider and observe carefully what happens when the shrews meet.

In new surroundings both shrews react rather similarly. They either stop and silently run away from each other or, if they bump into each other, one or both normally give a loud squeak and they spring apart. They do not seem to notice each other until they more or less run into each other.

A different type of response is found where one shrew has been living in the observation tank for some time. This is worth preparing in advance if it is planned to investigate fighting fully. A stranger is then introduced for observation as before. Crowcroft (1957) has summarized the possible responses neatly as:

1. Stranger leaps away before its presence has been detected, or when the resident freezes
2. The resident freezes, raises its muzzle as if to squeak, whereupon the stranger runs away
3. Both freeze, the resident screams and the stranger runs away.
4. The resident bites the tail of the stranger, which runs away.

It is not so easy to time the fights between two shrews which are both resident in a larger tank so that students can observe them in class time. But if students are to tackle such shrew interactions as a project, this situation does produce other and perhaps more dramatic fighting patterns. Contacts between two residents are characterized by screaming matches, mouth-to-mouth fighting, tail biting, and the most peculiar screaming frenzy where one or even both tiny animals throw themselves onto their backs in an apparent tantrum of screaming. Shrews rarely chase each other after a screaming or tail biting or tantrum throwing contest. However, it is common for them to dig furiously, possibly as a displacement activity after a battle. Students will also note that tail lashing is often a feature when the shrews are excited.

Time

30 minutes upwards

REFERENCES

Corbet, G.B. (1964), *Identification of British Mammals,* British Museum (Natural History), London.
Crowcroft, P. (1957), *The Life of the Shrew,* Reinhardt, London.
Van den Brink, F.H. (1973), *A Field Guide to the Mammals of Britain and Europe,* (3rd ed.) Collins, London.

24. Gerbils

INTRODUCTION

Gerbils are small elegant cricetid rodents originating from the scrub and desert regions of Southern Russia, Asia and Africa. The gerbil commonly used in laboratories today is *Meriones unguiculatus,* the Mongolian gerbil, which inhabits the semi-arid and desert regions of Mongolia and Northern China. This species was brought to Britain from China in 1964 and immediately found favour both as a pet and as a laboratory animal.

The Mongolian gerbil has several features which make it a particularly useful animal for behavioural and general studies from University to school level: they rarely bite; they are active during day time, they are not shy; they can be kept in pairs in small colonies with ease in the simplest of housing; they rarely eat their young, which means that with care the young can be handled a few days from birth; and they are relatively clean and odourless.

Gerbils are best kept in pairs in aquarium tanks covered with a wire-mesh lid. The floor of the tank can be covered with wood chips, sawdust, paper or other absorbent material. Our best colonies thrive in tanks floored with a 50:50 mixture of coarse sand and garden soil to a depth of 5 cm. This seems to keep the animals' coat and claws in good condition. Each tank should also contain a piece of flat stone, a piece of hard wood, a standard drip drinking bottle, and a supply of sacking or cotton waste for nesting materials. The diet is similar to that of other rodents, but care should be taken to include quantities of seeds, particularly sunflower and ground nuts.

Litters of 4 or 5 young may be produced fairly regularly every 30 or 40 days. The gestation period is approximately 25 days, and the young are normally weaned at about three weeks. They are sexually mature at about 14 weeks, and the average female will have 8-10 litters in her lifetime.

EXPERIMENTAL WORK

24.1. Investigation of paper-shredding

Purpose

A characteristic behaviour of the gerbil, of which many people will have had direct and bitter experience, is the animal's tendency to shred almost any material into small pieces. Since this characteristic shredding behaviour is easily observed and measured in a laboratory situation, it makes a particularly suitable topic for study.

A number of hypotheses may be put forward to account for the function of shredding behaviour:

(a) It is feeding.
(b) It is to build a nest in which to be warm and comfortable.
(c) It is shown by females to build a nest for the young.
(d) It is to keep the incisor teeth in good condition.
(e) It is abnormal behaviour induced by boredom or lack of exercise.

These hypotheses can be presented to students before the start of the study or students can be asked to draw up a list of possible functions after a preliminary period of observation.

Preparation
None

Materials

Per pair of students:
Tests of hypothesis (*a*)-(*e*) require the following items. Requirements for individual tests are indicated in the Table which follows.

1. 4-8 male gerbils separately caged
2. 4-8 female gerbils separately caged
3. 2 stop watches, one of which should be able to stop and start without returning to zero
4. A mercury thermometer
5. A student monocular microscope (or equivalent) with × 40 objective
6. A plastic tray on which to tip home cage contents
7. 16 unpainted soft wood blocks, 5 cm × 2 cm × 2 cm
8. 2 aquarium tanks 40 cm × 25 cm × 25 cm or similar-sized container for housing gerbils
9. 2 activity wheels
10. Plenty of pieces of paper 15 cm × 10 cm
11. 2 Pasteur pipettes with rubber bulbs
12. A Bunsen burner.

Hypotheses	Items
a	1, 6, 10
b and *c*	1, 2, 3, 4, 5, 6, 10, 11, 12
d	1, 3, 6, 7, 10.
e	1, 2, 3, 6, 8, 9, 10.

Per class:
1. A 10°C 'cold' room or equivalent situation
2. A 30°C 'warm' room or equivalent situation
3. A bucket of clean dry sand
4. A balance for weighing wood blocks
5. Materials necessary for staining vaginal smear slides in aqueous eosin and Erlich's haematoxylin.

Methods

(*a*) *To test whether paper shredding is a form of feeding.*
If a gerbil has the contents of its cage emptied out and is left for 5 minutes in the empty cage then, when a piece of crumpled paper is placed in the cage, paper shredding should begin quite quickly. This should be repeated 3 or 4 times, and careful observations made. It will be seen that the paper is held in the front feet and fragments chewed from it. Individual fragments may be separated from the sheet by an upward jerk of the head which is not seen in feeding. It will also be noticed that the paper fragments are not eaten but just dropped on to the floor. These features suggest that paper shredding is not feeding behaviour.

(*b*) and (*c*) *To test whether paper shredding is to produce a nest, either to keep warm and comfortable, or built by females only to provide a suitable environment for their young.*

Observations will quickly reveal that not only do female gerbils shred paper whether pregnant or not, but that males also show the same behaviour. Paper shredding is therefore not particularly associated with pregnancy in females, nor is it an exclusively female behaviour. This, however, leaves to be investigated the question of whether paper shredding is associated with the animal's lack of a nest, and therefore serves the function of producing nest materials. If this hypothesis is true, then more paper shredding should be expected when little or no nest is in the home cage, compared with when plenty of nest material is present.

The association between paper shredding and the possession of a nest may be simply tested by removing all nest materials from the home cage for 5 minutes before introducing a crumpled piece of paper, about 15 cm × 10 cm. The number of seconds spent shredding the paper during a 5-minute test is then measured, using one stop watch to measure the duration of the test, and a second one to add together the shredding bouts within the test. Control tests should also be carried out in which the nest materials are emptied out of the cage, then immediately reintroduced, thereby providing the disturbance but leaving the animal with its nest. After a 5-minute wait to allow the animal to settle, the paper shredding test should be carried out as for the experimental situation. Each animal may be used as its own control, half receiving the control test first and half the experimental test. This should be carried out on at least four animals. This exercise therefore introduces ideas of experimental design to control possible confounding variables, as well as seeking to answer a particular behavioural question.

The results will, none the less, probably be unclear. In the first place there will probably be appreciable variability of paper chewing in both experimental and control groups and, secondly, quite substantial amounts of paper chewing will be seen even when the nest is present, and no shredding may sometimes occur, even when the nest has been removed. The evidence does not therefore support the suggestion that paper shredding serves the sole purpose of producing suitable nest material.

An extension to this investigation is to see if paper shredding is correlated with environmental temperature: more occurring when the temperature is low, to produce a larger nest to keep the animal warm. Gerbils to be tested should be maintained between tests at a room temperature of about 20°C. Prior to a test, a gerbil should have its home cage contents removed and then be left for about 20 minutes at one of the three temperatures 10°C, 20°C, or 30°C. It is possible to improvise a bit in devising these temperature situations, provided the class numbers are small; room temperature gives the 20°C situation; above a radiator, 30°C, and an unheated room somewhere else, 10°C. This will raise problems of the effects of other differences between the test situations (such as noise), but this is not a serious objection, provided that students are aware of the limitations of the experiment. After the 20 minutes for the test temperature to take effect, a crumpled piece of paper, 15 cm × 10 cm, should be placed in the empty cage and the number of seconds of paper shredding in a 5-minute test measured as before. Each animal should be tested at all three temperatures.

Results will probably show that a considerable amount of paper shredding may be shown at all three temperatures (50-90% of test time spent shredding), but there is some evidence from pilot experiments that, at least for males, the correlation between shredding and temperature is the reverse of that predicted, more occurring at the higher temperature. This effect is not apparent for females.

It is possible that, in the case of females, paper shredding varies with the state of oestrus, since the oestrous cycle in rats is known to affect behaviour measures such as wheel running activity (see figure 25.2.1 on page 166). This could be followed by conducting daily paper shredding tests on female gerbils followed by a vaginal smear (see Section 25.3) to establish the state of oestrus. The oestrous cycle in the gerbil is about 12 days, so each animal would need to be followed over about a 2-week period.

(d) To test whether shredding behaviour serves to keep the incisor teeth in good condition.
At the start of the experiment, two groups of about four animals should be set up in their individual cages; the control group with only paper in the cage, which is not replaced after it has been shredded, and the experimental group with a block of wood 6 cm × 2 cm × 2 cm in the cage in addition to the paper. Both groups should be given some kind of soft food which will not significantly wear down the teeth.

The two groups of animals should be left for about a week to allow some incisor teeth growth before all animals are given a 5-minute paper-shredding test. This test should be repeated on the following 2 or 3 days, the prediction being that more shredding should be shown by animals without wood blocks, because their teeth are more in need of wearing down.

The hypothesis would also predict that, if wood blocks were given to animals deprived of the opportunity to chew on something hard for a week, they would show more wood chewing than animals with wood constantly available. This may be tested by providing all animals of both groups with fresh weighed wood blocks, and then reweighing the blocks after 24 hours to determine the amount of wood chewed away.

Observations suggest that appreciable amounts of paper and wood chewing may be expected from both groups, and that no clear difference between them will be apparent. This suggests that condition of the teeth is not a major factor in provoking chewing behaviour.

(e) To test whether paper shredding is due to boredom
A group of about four control animals should be housed individually in small cages (about 35 cm × 15 cm × 10 cm) containing only paper and food. This is the 'bored' group. The experimental or 'interested' group should be housed in any situation which provides much more interest and variety than the controls. It could, for example, be an aquarium tank of about 40 cm × 25 cm × 25 cm containing pairs of gerbils or all four together to provide social stimulation.

Additional variety could be provided by covering the floor with a few centimetres depth of sand, by including ramps, shelves or tubular runways, or an activity wheel.

After a period of 24 hours to acclimatize, each animal in both groups should be given a paper-shredding test. This may be repeated for the next two days to provide a reasonable sample. The finding is, however, likely to show that appreciable paper shredding occurs in both groups,

suggesting that variety of habitat, at least over that period of time, has no effect on paper shredding.

The results of the investigations do not seem to provide a convincing confirmation of any of the original hypotheses. This does not mean that none of them is correct, since these results are based on fairly preliminary observations. It does, however, mean that no clear-cut result is likely to be obtained in the kind of class exercise or project which you are able to undertake. This does not weaken the value of the exercise, since it encourages the systematic investigation of each of a number of hypotheses, so that they may be confirmed or rejected. It also raises the important point that biological problems often do not have simple clear-cut answers. A discussion is probably advantageous to put over the points that, firstly, there may be a function to the behaviour other than one of those hypothesized, but that unfamiliarity with the animal causes us to overlook it; and, secondly, that limitations of the experiments may have prevented an effect from being shown.

In summary it can reasonably be said that paper shredding looks as though it has some connection with nest building, since many of the fragments are usually incorporated into a nest. If this is true, then it may be that the frantic paper shredding shown by gerbils is a pathological form of nest building induced by long-term exposure to laboratory conditions; however the gerbils do in other respects appear to behave 'normally'.

Time
(a) This should take 10-30 minutes of observation.
(b) and *(c)* Investigating the effect of the presence or absence of the nest on paper shredding will take $2\frac{1}{2}$-3 hours. Experiments on the influence of temperature will require 2-3 hours for 3 days. Examining the effect of oestrous cycle will require 2 hours a day for 2 weeks, so as to follow all females through one complete cycle.
(d) This requires to be set up for one week before any investigations. After this, 2-3 hours of work will be required for 4 days.
(e) This requires 2-3 hours on the first day to set the experiment up, followed by 2-3 hours of investigation time for the next 3 days.

24.2. Observations on nest construction

Purpose
To make observations on the selection and modification of materials for a nest and their incorporation into the next structure.

Preparation
Dyeing cotton wool in two or three different colours

and drying them thoroughly before the observations start.

Materials

> Per pair of students:
> 1. 2 gerbils (male or female) housed in separate cages
> 2. An ample supply of nest building materials in the form of
>   (*a*) Pieces of paper of 3 or 4 different colours
>   (*b*) Pieces of card of 3 or 4 different colours
>   (*c*) Cotton wool of 3 or 4 different colours.

Methods

At the start of observations all nest materials should be removed from the home cage of a gerbil, and it should then be given one type of material (or a choice of materials such as paper and cotton wool) from which to make a nest. These materials should be left till the following day, when notes should be made on the proportions of materials incorporated into the nest and whether different materials are used in different parts; this should cause as little disturbance to the nest as possible. After notes have been made, any unshredded materials provided the day before should be removed and new materials, of a different colour, introduced. The gerbil should then be left till the following day, when the same process may be repeated. This procedure can be carried on over four or five days.

In general it will be found that paper, card and cotton wool will all be incorporated into the nest, after first being shredded into small fragments; they will not be incorporated into different parts of the nest, but all mixed up together. Materials added on different days will also be found to be mixed up together, indicating that nest materials are frequently re-arranged.

These studies may be made more detailed by, for example, weighing materials before introduction and the unshredded pieces removed the following day to determine weights of materials incorporated each day. The shape and size of fragments in the nest may be assessed from nest samples taken each day. This may well show that the mean size of nest fragments decreases over the days, showing that shredding of nest fragments is a continuing process.

The results raise some useful questions on the interpretation of results from unnatural situations. The homogeneous mixing of the different nest materials may be substantially effected by digging behaviour, which may be functionally unrelated to nest construction. The continued shredding of nest fragments raises problems already dealt with in Section 24.1.

Time

15-30 minutes a day for 4 or 5 days would produce worth-while results. More time per day would be required if nest materials are to be weighed and measured.

24.3. Observations on burrowing

Purpose

If gerbils are kept in small cages, even with plenty of nest material, it will be observed that they show frenzied bouts of scraping at the sides and in the corners of the box as if trying to get out. Whether or not they are trying to escape, the observation does suggest that gerbils are active diggers. This exercise allows gerbil digging behaviour to be observed.

Preparation

Enough clean dry sand should be prepared to $\frac{2}{3}$ fill an average aquarium tank.

Materials

> Per class or student group:
> 1. 1 or 2 adult gerbils.
> 2. 1 aquarium tank with glass or transparent plastic sides and a roof.
> 3. Enough sand to $\frac{2}{3}$ fill the tank.

Methods

The tank should be fitted up with the sand, and then pressed down a bit to make it fairly firm. One or two gerbils should then be introduced into the tank. If more than one animal is to be introduced, it is better that they should already be familiar with one another, otherwise they may well fight. It is, however, possible to introduce two strange animals without severe fighting, provided that they are both put into the tank at the same time. Swanson (1974) also shows that fights are less likely to occur when two strangers of the same sex meet, and suggests that fights may not be so severe between animals that are not yet fully grown.

Newly introduced gerbils, apart from social interactions, will at first show a lot of running about and scent marking with their ventral gland (see Section 24.4). Soon however small 'test' cavities will be dug at the sides and in the corners of the tank, with rapid alternating movements of the front legs. This will change to rapid determined digging, where piles of sand being excavated by the front feet will be vigorously kicked away by the back feet. In this way a steep burrow or perhaps a wide crater is dug. This will then be extended with subterranean burrows whose progress along the floor and sides of the tank can still be seen through the glass walls.

Paper or cotton wool may be provided at this stage; by the following day it will have been taken down to an underground nest cavity as nest material. If the burrow system is observed for the next few days it will be found, as in the case of the shrews (Section 23.1), to be constantly changing, although this may in part be due to burrows collapsing.

These observations will emphasize the adaptiveness of the gerbil to underground life – which cannot be appreciated if the gerbils are kept in the usual laboratory cages.

Time

It will take an hour or so from the time of introduction for gerbils to have accomplished any extensive digging. This can, therefore, be set up as a class demonstration which can run while students are chiefly engaged in some other work, but can spare 2-3 minutes at intervals to observe the digging. The demonstration may then be left for several days, so that the construction of a nest cavity and the daily alteration in the burrows system may be noted.

24.4. Demonstration of scent marking

Purpose

Both male and female gerbils have modified sebaceous glands on their chests. This chest area produces a secretion which is rubbed on the substrate by a characteristic scent-marking movement (figure 24.4.1). The purpose of this exercise is to demonstrate scent marking behaviour.

Preparation

None

Figure 24.4.1 Male gerbil marking by dragging the ventral sebaceous gland along the substrate.

Materials

1. 1 adult male gerbil in its home cage
2. 1 plastic tray on which to tip home cage contents.

Methods

If a male gerbil is removed from its home cage, and the contents of the cage are tipped onto a tray, the gerbil, when returned to the home cage, will immediately show repeated dragging of its ventral region along the floor as it investigates all parts of the cage.

Time

5 minutes.

24.5. Investigation of scent-marking behaviour in males and females

Purpose

To provide a simple experimental situation where scent marking may be investigated.

Materials

Per pair of students:
1. 3-6 adult male gerbils in individual cages
2. 3-6 adult female gerbils in individual cages
3. 1 transparent moulded plastic aquarium tank about 50 cm × 30 cm × 30 cm (glass aquarium tank will do).
4. 3 glass bottle stoppers to act as scent marking pegs.
5. 1 small bottle of dilute acetic acid.
6. 1 stop watch or clock
7. Cotton wool.
8. Some Plasticine.

Methods

Before any animal is tested, a single glass stopper should be stuck down in the centre of the floor on a small piece of Plasticine. One gerbil should then be placed in the tank and observed during a 5 or 10 minute period, and the following scored:

(a) Number of times the peg is scent-marked.
(b) Number of times the floor is marked immediately beside the wall (the edge 3 cm).
(c) Number of times the rest of the floor is marked.

After the test the stopper and its Plasticine should be removed, the floor and all the sides wiped thoroughly with cotton wool soaked in dilute acetic acid, and then wiped dry. This should remove traces of marking by the gerbil. A clean peg should then be fastened to the centre of the tank with a fresh piece of Plasticine and the test repeated. After all the animals have been tested, it should be fairly apparent that males scent-mark more than females. It will probably also be found that the peg is the most frequently marked of the three areas chosen, with the edges of the floor the next most frequent. All these results can be most clearly demonstrated if there is an opportunity for pooling class results. Thiessen, Blum and Lindzey (1969) using a somewhat more sophisticated apparatus also found greater marking frequency in males. This corresponds to the larger size of the scent gland in males compared to females. However, they point out that there are a number of explanations possible for the function of the behaviour. They give the following:

(a) To aggregate members of the species.
(b) To disperse members of the species.
(c) To attract members of the opposite sex.
(d) To allow individual recognition.
(e) To mark individual or group territories.

The problem may be investigated further with the same apparatus as above by observing the response of a gerbil (male or female) to a peg which has already been marked by another individual (male or female). Thiessen, Lindzey, Blum and Wallace (1970), using a Y-maze situation, found that male gerbils were more attracted to the arm of the maze with male scent than the one with female scent. Females, however, showed no preference. However, we have seen some tentative evidence from a student study on the Lybian gerbil that, in a situation with an arena and a marked peg, both males and females sniffed and marked a peg with female scent more than one with male scent. This is worth looking at again. Thiessen *et al.* (1970) suggest that the attraction of males to male scent indicates a territorial function in male scent marking, males wishing to 'overmark' rival males' sebum. This does not mean that scent marking may not have other

functions, even in males. Thiessen *et al.* (1970) also found a positive correlation between the amount of scent marking in an arena and the dominance shown in male-male pair encounters, such as is described in Section 24.6. This would allow the two sections to be linked in a project type of exercise.

Time
If carefully recorded data are needed, then to test 3 or 4 males and the same number of females in the clean peg situation would take about 2 hours. A further 3 or 4 hours would be needed to test a similar number of males and females with male and female marked pegs. For larger classes it may be possible to reduce this time appreciably, by getting each pair of students to test only one or two animals in each situation and pooling the class results.

24.6. Behaviour shown to a stranger

Purpose
To investigate the differences between male and female gerbils in the behaviour shown when encountering a stranger of the same or opposite sex.

Preparation
Swanson (1974) in a similar study suggests that animals of 5 to 6 months should be used. (That is an age when they are sexually mature but not full adult weight.) The reason is that fighting between strange animals at this age is less savage and is unlikely to result in injury. If enough animals of the right age are to be obtained, some prior planning may be necessary.

Materials

Per pair of students:
1. 3-6 adult male gerbils in individual cages.
2. 3-6 adult female gerbils in individual cages.
3. 1 aquarium tank about 50 cm × 30 cm × 30 cm with a wire net lid.
4. 1 stop watch or clock.
5. Half a bucket of sawdust.
6. Magic Markers of 2 or 3 colours.
7. 1 cassette tape recorder or score sheet.

Methods
Before any testing, all gerbils should be individually marked with conspicuous coloured spots on the back

or tail with Magic Marker. Before each test the floor of the tank should be covered with fresh sawdust to allow the gerbils to grip. Pairs of animals should be introduced simultaneously into the tank and the lid replaced. The behaviour shown by each animal should then be observed and recorded over a 5 or 10 minute test period. The behaviours which may be scored are:

(a) Sniffing (scent-gland or genital area) (figure 24.6.1)
(b) Upright threat (figure 24.6.2)
(c) Sideways threat (figure 24.6.3)
(d) Attack – one individual upon another.
(e) Fight – mutual wrestling and biting.
(f) Chase.
(g) Drumming ground with hind feet.
(h) Scent marking (figure 24.4.1).

Figure 24.6.1 Sniffing the scent gland of a strange gerbil.

Figure 24.6.2 The gerbil on the left leans forward, ears forward and neck fur raised, in the upright threat position. The gerbil on the right beats a rapid retreat.

Scoring all these will be possible with a tape recorder but, with a score sheet alone, only two or three will be possible for an inexperienced observer. The sort of results that should emerge are that nosing, sniffing and threatening are quite common in single-sex encounters, and these may lead to fights or chases; however fighting and chasing are more likely to be shown in male-female encounters than single-sex ones. Females almost always initiate the fight by attacking the male, and frequently end up chasing him.

Swanson (1974) found that males marked more than females, and marked more in the presence of

Figure 24.6.3 The nearest gerbil arches its back and pushes against its rival with sideways threat.

females than in the presence of males. Males also differed from females in showing drumming behaviour, although it is not very clear what the function of this behaviour is.

These results show that there are differences shown by the sexes in their behaviour towards strangers, and that this is dependent on social context. They also show that, contrary to the expectation of some students, females are capable of showing vigorous aggressive behaviour.

Time
This is best suited to a project type of study, because scoring of the behaviours will require some familiarity with the animals, and taking the results off a tape requires at least as much time as it takes to record them; however, two hours a day for a week would give useful results.

REFERENCES

Swanson, Heidi H. (1974), Sex Differences in Behaviour of the Mongolian Gerbil (*Meriones unguiculatus*) in Encounters between Pairs of Same or Opposite Sex, *Anim. Behav.*, 22, 638-644
Thiessen, D. D., Blum, S. L. and Lindzey, G. (1969), 'A Scent Marking Response Associated with the Ventral Sebaceous Gland of the Mongolian Gerbil (*Meriones unguiculatus*),' *Anim. Behav.*, 18, 26-30.
Thiessen, D. D., Lindzey, G., Blum, S. L. and Wallace, P. (1970), 'Social Interactions and Scent Marking in the Mongolian Gerbil (*Meriones unguiculatus*),' *Anim. Behav.*, 19: 505-513.

25 Mice and Rats

The mouse and the rat are well-established laboratory animals, and this Section seeks to exploit a readily available source for simple exercises in rodent behaviour. For this purpose it is convenient to treat them together.

Colonies of albino mice and other variants are frequently available for genetic or physiological work. These, plus colonies of the wild House Mouse *Mus musculus*, should be considered for possible behaviour work. The house mouse, originally mainly Palaearctic in distribution, is now world-wide. The species tends to be divided into the shorter-tailed light-bellied forms living wild, and the longer-tailed grey-bellied forms living commensally with man. A most readable account of the behaviour of the wild mouse is given in Peter Crowcroft's *Mice All Over* (1966).

Similarly, the laboratory and wild strains of the rat *Rattus norvegicus* can be used for ethological exercises.

The genus *Rattus* is large with over 550 named forms distributed all over the Old World. A few species have become most successful commensals of man, and the distribution of the Brown Rat *R. norvegicus* is astonishing in its versatility, reaching S. Georgia in Antarctica and Spitzbergen and Alaska in the Arctic. *R. norvegicus* is still spreading in the New World at the expense of the Black Rat *R. rattus*. A good account of the behaviour of the wild brown rat is to be found in S. A. Barnett's *A Study in Behaviour* (1963).

Colonies of mice and rats are, given space, reasonably easy to set up and maintain. Biological suppliers will provide all necessary wire caging, food hoppers, drip bottles and standard diets in pellet form.

25.1. Rat: A simple measure of dominance

Purpose

Adult male rats show certain characteristic postures to indicate threat and appeasement (Barnett, 1963; Grant and Mackintosh, 1963); however a familiar group of male rats show little or no aggression towards one another in the normal way. This exercise provides a way of obtaining an estimate of dominance in a situation where hungry rats compete for food.

Preparation

A group of 4 to 6 young rats should be placed together 2 or 3 days before the class to allow them to establish stable dominance relationships. For 12-24 hours before the class the food but *not the water* should be withdrawn.

Materials

Per pair of students:
1. 4-6 young male rats (6-8 weeks old) in home cage
2. 1 spare cage
3. 1 stop watch
4. Some pellets of rat cake.

Per class:
1. Magic Markers of assorted colours.

Methods

All the rats should be individually marked on the tail with Magic Marker. All the rats except two should then be removed from the home to the spare cage and a single pellet of rat cake placed in the food hopper of the home cage. The two rats will attempt to feed from the pellet, pushing each other away with the front feet.

165

The rats should be observed for 1-2 minutes, and the number of pushes delivered by each of the rats counted. This gives a measure of dominance. Each rat should be tested with all others to establish a rank order. Rats will manage to eat a little during the test, which will lower their inclination to compete; each rat should therefore be tested more than once with the same opponent on separate days to give a more accurate estimate.

Time

2 hours a day for 2-4 days. As a simple demonstration of dominance, 5 minutes.

25.2. Mouse or Rat: Open-field behaviour

Purpose

The behaviour of small rodents in an arena or 'open field' has been a popular measure of 'emotionality' in experimental psychology. The basic assumption has been that a more-fearful animal will show less locomotion and more defaecation than a less-fearful or emotional one. This assumption is not without critics, and reviews of the value of open-field tests may be found in Denenberg (1969) and Archer (1973). The great merit of the test is its simplicity, and the results do form a useful basis for discussions of the interpretation of behaviour. These results may also be used in a study comparing open-field performance with learning ability in 25.4.

Preparation

It is necessary to build open fields as required. The specifications are not very rigorous. The walls should be high enough to stop animals jumping out. The arena may be round or square, and its floor surface should be divided into parts of equal area. The total floor area should be large enough to allow an animal to run around; a square surface 75 cm × 75 cm divided up into 15-cm squares would be reasonable for mice or rats. The inner surface of the arena should be easy to wash clean of faeces and urine.

Materials

Per pair of students:
1. 4-6 male or female rats or mice (about 8 weeks old)
2. An open-field arena
3. A stop watch.

Methods

Place one animal in a central square of the arena and observe for 5-10 minutes, scoring the following behaviour:

  (*a*) Number of lines crossed
  (*b*) Number of 'edge' squares entered and number of 'centre' squares entered.
  (*c*) Number of times the animal reared up against the wall as if trying to get out.
  (*d*) Number of times the animal reared up but not against the wall.
  (*e*) Number of faecal pellets.
  (*f*) Number of urine spots.

The expectation is that a less-fearful rat will have a high activity score, and will spend a fair proportion of its time in squares away from the edge of the arena; it will explore the arena, rearing up against the wall or in the centre. A more-fearful animal will keep to the edge of the arena, show low locomotion and rearing scores, and higher defaecation and urination. Broadly speaking this appears to be true for rats and mice, but the expected correlations between the different measures may not be found, and so students should be encouraged not to pool scores of the different measures. If male rats or mice are being used, the performance of the animals in the open field may be compared with their learning ability in a Skinner box (25.4). For females, the performance in the open field may be examined in relation to the oestrous cycle (25.3) since activity in rats and mice is known to fluctuate with phase of the oestrus cycle (figure 25.2.1).

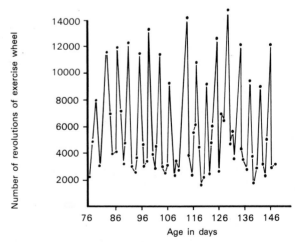

Figure 25.2.1 The four-day cycle of activity in the female rat corresponding to the oestrous cycle (after Richter, 1952, from Barnett, 1963).

Time

To get a reasonable amount of data, all animals should be tested 2 or 3 times, preferably on separate days. This will take about 1 hour each day.

25.3. Mouse or Rat: Determination of oestrous cycle

Purpose

The activity, aggressiveness and other means of behaviour in adult female rats and mice shows fluctuations in synchrony with the oestrous cycle (figure 25.2.1). The length of the cycle in the laboratory mouse is about 4 days and in the rat 4-5 days; this means that a fairly short project study may examine at least one cycle in relation to some measure of behaviour such as open-field performance. This exercise describes how oestrous state may be determined from a stained vaginal smear.

Preparation

None

Materials

Per pair of students:
1. 4-6 adult female mice or rats
2. 2 fine Pasteur pipettes with bulbs
3. Box of microscope slides
4. Box of coverslips
5. Staining dish of 70% alcohol
6. Staining dish of Ehrlich's haematoxylin
7. Staining dish of alkaline water
8. Staining dish of acid water
9. Staining dish of aqueous eosin
10. Canada balsam
11. Distilled water
12. Wire loop
13. Stop watch
14. Bunsen burner
15. Microscope with × 40 objective
16. Magic Markers.

Methods

First the Pasteur pipettes should be lightly flamed at the top; this removes any roughness in the glass and narrows the opening. All animals should then be individually marked on the tail with Magic Marker.

To take a vaginal sample, a mouse should be placed on the top of its cage so that it can hold onto the wire and be held by the tail. The tip of a Pasteur pipette containing a small drop of water should be inserted into the vagina, and the water expelled and drawn in again, bringing with it cells from the vagina in suspension.

The drop of liquid taken from the vagina should be placed on a clean slide and spread over it with a wire loop. The slide should then be dried over a low Bunsen flame to fix the cells to the slide. The slide may be stained in a standard manner, using haematoxylin and eosin. The smears should then be covered with Canada balsam and a cover slip, and examined under the microscope. The identification of the phases of oestrus is as follows:

1. Prooestrus: Phase of active growth of the genital tract. Leucocytes and nucleated epithelial cells are seen in the smear in almost equal numbers. A few non-nucleated (cornified) epithelial cells may also be present (figure 25.3.1).

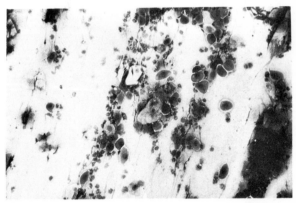

Figure 25.3.1 *Prooestrus* (rat) – showing presence of leucocytes (small nucleated cells), and nucleated epithelial cells (large rounded cells) in almost equal numbers.

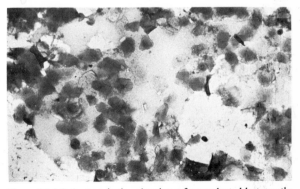

Figure 25.3.2 *Oestrus* (rat) – showing a few nucleated but mostly cornified (non-nucleated) epithelial cells with leucocytes almost absent.

H

2. Oestrus: Stabilization of genital tract growth. Nucleated and cornified epithelial cells may both be present, but no leucocytes (figure 25.3.2)
3. Metoestrus: Degenerative phase of the genital tract. This is characterized by cornified epithelium and leucocytes in the smear (figure 25.3.3)
4. Dioestrus: Resting phase of growth of genital tract. The smear showing almost exclusively leucocytes with some nucleated epidermal cells (figure 25.3.4).

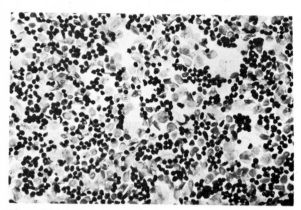

Figure 25.3.3 *Metoestrus* (rat) – showing a mixture of cornified epithelial cells and leucocytes.

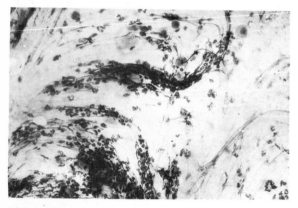

Figure 25.3.4 *Dioestrus* (rat) – showing almost exclusively leucocytes in a viscous mucus.

Time

1-1½ hours to take samples from 6 mice, stain and examine them. For an extended study this will need to be carried out on the same mice over the duration of at least one cycle of 4 days.

25.4. Rat: Learning in a Skinner box

Purpose

The Skinner box is a device in which an animal learns to perform an operation, usually pressing a lever, in order to obtain a food reward. This situation has been much used by experimental psychologists to study parameters of learning, firstly, because it provides a standard experimental situation and, secondly, because it is rather versatile.

The Skinner box has the disadvantage from the point of view of the student of animal behaviour in being a very unnatural experimental situation; however, it does allow many features of learning to be observed and studied. It has the practical disadvantage of being an expensive piece of apparatus if obtained commercially, but much less expensive ones may be devised and built if you should have some workshop facilities of your own. A design, which we have not had the opportunity to try ourselves, but has been recommended to us as cheap to build and run, as well as reliable, is one devised originally by Deutsch which dispenses cut lengths of spaghetti.

By pressing a lever the rat operates a microswitch; this completes a circuit through a solenoid, which draws back a brass knife allowing a length of spaghetti to fall down onto a platform. When the circuit is broken, the knife returns under the tension of a spring and cuts off a predetermined length of spaghetti. The cut length of spaghetti slides down a chute into a food dish (figure 25.4.1*a*).

A simple design for liquid reward recommended to us is a solenoid-operated dipper which delivers the liquid in a cup through a small hole in the floor (figure 25.4.1*b*).

Preparation

Rats should be deprived of food, but not water, 12-24 hours before testing.

Materials

Per pair of students:
1. 6 male rats, about 8 weeks old
2. 1 Skinner box
3. Supply of Skinner box pellets
4. Stop watch or clock.

Per class:
1. One spare Skinner box for every 6 or 8 students to allow for mechanical faults
2. Magic Markers in assorted colours.

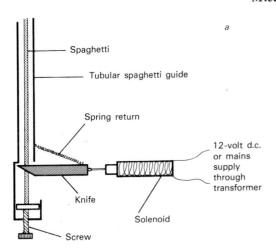

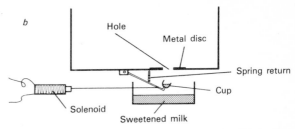

Figure 25.4.1 (*a*) Diagrammatic representation of a Skinner box system to dispense cut lengths of spaghetti. A platform below the knife can be adjusted by means of the screw indicated to provide different sizes of reward.

(*b*) Diagrammatic representation of a simple Skinner box system for dispensing liquid reward. The solenoid pulls the dipper down into the liquid. The spring returns the filled cup to floor level. The hole should be surrounded by a metal disc on the floor of the box to indicate the position of the hole to the rats.

Methods

(a) Acquisition of the response

At the start of the training each rat should be individually marked on the tail with Magic Marker; each should then be placed in turn in the Skinner box for 10-15 minutes. The training of the rat should then be begun by *shaping* the response. This involves the gradual leading of the rat to the full lever-pressing response by first rewarding it for an only partial response. So, at first, a pellet of food is constantly available in the dish, so that the rat discovers the location of food; when the rat has discovered this, a pellet will be delivered only when the rat approaches the dish; then, after a time, only when it touches the lever and, finally, it will press the lever to deliver a pellet to itself. The response will be learned gradually in a series of one or two trials a day over about 5-7 days. When the rat has learned the response, it may eat up

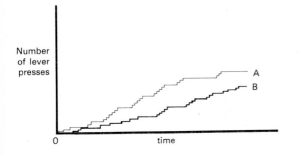

Figure 25.4.2 Cumulative record of lever presses for two rats A and B during a 10-minute trial. A steep rise in the line indicates a rapid lever-pressing rate, and a horizontal line indicates a pause in lever pressing.

to 100 pellets in a 15-minute trial. After the animals have had their trials for the day, they should be given free access to food for about an hour before it is taken away again.

During the training, a record should be kept of the number of pellets given to each rat in each trial, and under what circumstances. Some students may have problems in training their animals; this is often due, not to be stupidity of the rats but to poor shaping or rough handling. The moral is that rats are more sensitive creatures than they appear to be.

When lever pressing is well developed, it is convenient to record performance on the cumulative lever-pressing record over time (figure 25.4.2); a steep line indicates a rapid lever-pressing rate, and a horizontal line, a pause.

(b) The influence of shaping

Rats will show individual variation, not only in their rate of acquisition, but in the behaviour they show accompanying acquisition. One rat may learn to press the lever with his left foot, another with his teeth. Some rats may show behaviour patterns in association with lever pressing, such as bobbing up and down or wiping the face. These can be used to demonstrate the power of shaping the response.

As soon as the rat has achieved a constant level of lever-pressing performance, the base-line occurrence of some behaviour pattern, say bobbing up, should be scored in two trials on one day. After that, a shaping procedure should be adopted to reinforce bobbing up in an ever-exaggerated form. At the end of a series of trials it may be possible to get a rat to rear right up to touch the roof of the box before a bar press. It should at least be possible to show an increase in the occurrence of the behaviour compared with that before shaping.

(c) Size of reward

A feature of learning is the correlation between size of reward and rate of acquisition of the response. It should be possible to demonstrate this using the spaghetti Skinner box (figure 25.4.1*a*) to dispense different lengths of spaghetti to different individuals.

(d) Secondary reinforcement

As soon as the basic lever-pressing response has been acquired, one can say that the pressing of the lever is the operant response to the *primary reinforcing stimulus* of food reward. It is now possible to demonstrate that any stimulus can take on the role of reinforcer by association with the primary reinforcer; i.e. an initially non-rewarding stimulus may become a *secondary reinforcer*.

Most commercial Skinner boxes have a light. After the initial acquisition of the lever pressing, the light should be paired with the presentation of the food reward. After this has been repeated for a few trials, both light and food should be withdrawn. After a while this will cause the rat's lever pressing almost to stop. If the light is again paired with lever pressing, the response rate of the rat will go up again, even though no food reward is delivered. This demonstrates that the light has become a *secondary reinforcer*. A useful practical point is to keep the Skinner box in subdued light, so the box light is more obvious to the rat.

(e) Contiguity

One of the important properties of reinforcement is that the operant response develops more readily the closer in time the reinforcing stimulus appears after the response; this phenomenon is called *contiguity*.

This may be demonstrated by observing the number of trials it takes to train a group of rats, each given a different delay time from bar pressing to presentation of reward varying from 0 to about 20 seconds.

(f) Partial reinforcement

It may reasonably be argued that in nature animals must seldom be rewarded every time for making a particular response, and yet nonetheless they seem to learn. This is a good reason for studying learning in a Skinner box situation, where reward is given according to some schedule which does not provide a reward for each response; this is called *partial reinforcement*. It is beyond the scope of this book to deal with the great variety of partial-reinforcement schedules, but here are the most common and simple ones, each of which produces a characteristic response from the rat:

> *Fixed ratio* – Reinforcement is given after a fixed number of responses; say – every 5th lever press. This tends to produce very high rates of responding, even up to quite unfavourable ratios (1 in 10, or more). An animal trained on continuous reinforcement may be changed to increasingly unfavourable partial-reinforcement schedules.
>
> *Fixed intervals* – Reward is followed by a fixed interval of non-reward (say 1 minute). The first response after that 1 minute is again rewarded followed by another minute of non-reward. This is more difficult to establish than with fixed ratio but, on a 1-minute schedule, some evidence of time-keeping by the rat should become apparent, with the bar-pressing rate showing a clear rise towards the end of the 1-minute period.

A variation on this type of schedule is to take a rat just trained on continuous reinforcement, then train it to bar-press only when the overhead light is on, by providing 1 minute of continuous light accompanied by reward, followed by 1 minute of no light and no reward. During the training trials the number of bar presses in rewarded and non-rewarded time should be recorded for evidence of improved performance.

(g) Discrimination learning

As mentioned already, the Skinner box is a rather versatile situation and may be modified to investigate a variety of phenomena. One modification of the basic Skinner-box situation is that to study discrimination learning where a rat learns to press a lever when presented with one stimulus, but not when presented with another; there may indeed be two levers so that, for example, in a visual-discrimination learning situation, a circular symbol might be placed over the rewarding lever and a cross over the non-rewarded one. The non-rewarding lever may actually provide negative reinforcement, such as a puff of air in the face.

This situation has been used to study parameters of learning, such as the ability to re-learn the problem after the positive and negative stimuli have been switched (reversal learning); it has also been used to investigate the limits of sensory discrimination of animals which show appropriate learning ability, e.g. what is the minimum difference in sound frequency or wave length of light that a pigeon, dog or whatever can still detect. The details of such experiments are really outside the scope of this book.

(h) Extinction of the response

When a rat has been trained to lever-press, its response to complete removal of the food reward may be studied. A rat should be placed in a Skinner box and rewarded in the normal way for the first minute of the trial; delivery of the food pellets should then be stopped. The immediate response of the rat is a greatly increased lever-pressing rate, which may persist for a minute or two with short pauses. Gradually the pauses become longer, and the lever-pressing bursts shorter. The actual point of *extinction* is up to the investigator, but a convenient criterion is at the end of a 3-minute pause since the last lever-press. The time from the withdrawal of the reward to the extinction point gives the extinction time (figure 25.4.3).

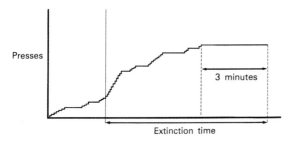

Figure 25.4.3 Measurement of extinction time: Withdrawal of reward results in a rapid rise in lever-pressing rate. This is followed by bursts of lever pressing which gradually become less frequent. After 3 minutes of no pressing, the trial is ended. This time from withdrawal of stimulus to the end of the trial gives the extinction time.

Students may predict that if the rat is tested again on the day following extinction, it will show no response if no food reward is provided; they will, however, find a partial *spontaneous recovery* which is nonetheless extinguished more rapidly than on the first day. Some spontaneous recovery may also be present on the third day of extinction. This can be used to bring out the point that an animal that has undergone extinction is not the same as an animal which never learned the problem.

(i) Comparison of Skinner-box learning with other behaviour measures.

During the training it will be found that different animals learn at different rates; this could be interpreted as showing that some animals are more intelligent than others; however, a more limited interpretation is that some rats are able to learn this type of problem more readily than others, but might not show the same superiority if tested in some other learning situation. To test this latter hypothesis, one student built a simple cardboard maze on a 50-cm-square ground plan with 7-cm wide corridors. He measured the number of blind-alley entrances and running time in the maze to obtain a food reward. He was able to obtain some correlation between performance by his 6 rats in the maze and their performance in a Skinner box, all in a three-week project. This suggested some general ability to learn – in this sample of rats at any rate.

Differences in learning rate in a Skinner box could, of course, be related not to intelligence but to some totally different parameter, such as fear of a strange situation or rank order within a group of rats. The first of these alternatives can be investigated by conducting the Skinner-box studies in conjunction with exercise 25.2 on performance in an open-field situation.

A method of measuring rank order within the group is described in exercise 25.1. Dominance might affect Skinner-box performance in two ways: Firstly, subordinate rats could just be more or less responsive than dominant individuals to the learning situation; secondly, subordinate rats could be more hungry because they are getting a smaller proportion of the limited food supply available in the competitive home-cage situation. An indication of the latter might be obtained by weighing all rats each day during their period of training.

Time

To test 6 rats once a day will take 2-2½ hours. If this is repeated 5 days a week, then all rats should be lever--pressing in 7 days. At least a further week will be needed to investigate the aspects of learning described in (b) to (i); two weeks or more being necessary for (e), (f) and (i).

25.5. Mouse or Rat: Maternal behaviour

Purpose

A feature of rats and mice is that the pups spend
much time hanging on to the nipple when the mother
is in the nest; if the mother leaves the nest rapidly, this
may result in pups being dragged from the nest attached to the nipple before falling off. It is probably
for this reason that female rats and mice show
well-developed retrieving behaviour in which the pup
is picked up in the mother's teeth and returned to the
nest (figure 25.5.1). This attractive parental behaviour
may be reliably demonstrated, and allows some scope
for experimental manipulation.

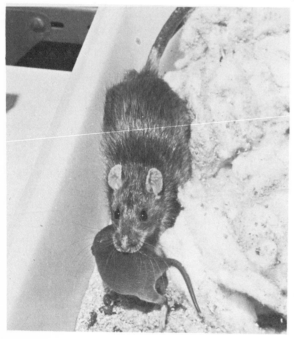

Figure 25.5.1 Mother rat retrieving a pup in her mouth and returning it to the nest.

Preparation

None

Materials

1. A cage containing rat or mouse parents with their litter
   of pups with eyes not yet opened
2. A second litter of pups of more or less the same age (optional).

Methods

If some of the pups are gently picked out of the nest
and scattered round the cage, the mother will become
quite excited and active; soon she will pick up a pup,
usually by the skin on the back, but sometimes by the
leg or tail, and return it to the nest. All the other pups
will be returned to the nest in a similar way. The pups
may struggle, but more often hang almost motionless
from the mother's mouth. After the last pup has been
retrieved, the mother will then explore the cage, apparently in search of any pups she may have missed.
It is known that rat and mouse pups which have fallen
out of the nest emit ultrasonic squeaks which attract
the mother (Sales and Pye, 1974). It might also be expected that a mother would respond to her own pups
only, since they probably have a characteristic smell;
however, it can be quickly shown that mothers will
retrieve pups not belonging to them, and mouse
mothers will even energetically retrieve young rat
pups substantially larger than their own; rat mothers
do not seem to respond so well to mouse pups.

Time

10-20 minutes

25.6. Mouse or Rat: Development of pups

Purpose

Rat and mouse pups change from their time of birth
as blind, naked and helpless objects to creatures like
small versions of their parents 3 weeks later. This
change in appearance is accompanied by corresponding changes in behaviour which may be observed.

Preparation

None

Materials

Per 4 students:
1. *Either* 1 litter of mouse pups in each of the following
   age ranges: 1st week, 2nd week.
   or 1 litter of rat pups in each of the following age
   ranges: 1st week, 2nd week, 3rd week.
A piece of stiff card about 20 cm × 20 cm covered in some
kind of cloth to provide an inclined surface up which pups
can crawl.

Methods

Pups should be looked at from day to day, and the

following features recorded for at least 2 pups in each age class:

Appearance
1. Whiskers present
2. Body fur present
3. Eyes open

Behaviour in the nest
4. Locomotion: movement of head and limbs
5. Grooming
6. Fighting, or rough and tumble

Behaviour out of nest
7. Righting response: turning the right way when the pup is placed on its back.
8. Locomotion: crawling with front legs or walking on all four.
9. Climbing: Ability to climb up a steeply inclined plane covered with cloth.
10. Response to edges: Does the pup walk over the edge of a table?

This will give the results shown in Table 25.6.1.

The overall difference between mouse and rat is the more rapid development of the former, this is particularly noticeable in the movement of the back legs, which appears in the mouse on day 2 but not in the rat till the middle of the second week. Some details of rat development may be found in Bolles and Woods (1964), but a day-to-day account of the development of several kinds of locomotor activity over the first 3 weeks is given by Altman and Sudarshan (1975).

Table 25.6.1 Guide to the results that may be expected from observations on the appearance of behaviour of mouse and rat pups from 1-3 weeks old.

MOUSE

	Week 1	Week 2
1 Whiskers	yes	yes
2 Body fur	no	yes
3 Eyes	closed	begin to open about day 10
4 Locomotion in nest	crawling and burrowing	active in nest
5 Grooming	no	yes
6 Fighting	no	yes
7 Righting	inefficient	instant
8 Locomotion	quite active, walking at end of week	rapid and efficient walking
9 Climbing	no – claws not developed	climbs up and down quite well
10 Response to edges	falls over edge	by the end of the week turns away from edges

RAT

	Week 1	Week 2	Week 3
1 Whiskers	yes	yes	yes
2 Body fur	no	soft velvet by end of week	yes
3 Eyes	closed	closed	begin to open about day 16
4 Locomotion in nest		active crawling and burrowing	active
5 Grooming		yes: face washing appears first, then scratch and lick	active
6 Fighting	no	no	fighting, rough and tumble, and exploration
7 Righting	inefficient	efficient	instant
8 Locomotion	weak crawling	walking – slow but steady	rapid efficient walking
9 Climbing	none	reasonable	climbs up and down
10 Response to edges	falls over edge	by end of week it hesitates at edges	turns away

Time

The observation of animals in the nest will require careful observation in a quiet situation for 10-30 minutes for each age class. For more introductory levels it may be preferable to leave this out. Observations out of the nest will be simpler and quicker; a total of 40-60 minutes would provide useful information on 2 rat pups of each of 3 age classes and 30-40 minutes to look at a total of 4 mouse pups.

REFERENCES

Altman, J. and Sudarshan, K. (1975), 'Postnatal Development of Locomotion in the Laboratory Rat,' *Anim. Behav.*, 23, 896-920.

Archer, J. (1973), 'Tests for Emotionality in Rats and Mice: A review,' *Anim. Behav.*, 21, 205-235

Barnett, S. A. (1963), *A Study in Behaviour*, Methuen, London.

Bolles, R. C. and Woods, P. J. (1964), 'The Ontogeny of Behaviour in the Albino Rat,' *Anim. Behav.*, 12, 427-441.

Crowcroft, P. (1966), *Mice All Over*, Foulis, London.

Denenberg, V. H. (1969), 'Open Field Behaviour in the Rat: What Does It Mean?' *ANYAS*, 159, 852-859.

Grant, E. C. and Mackintosh, J. H. (1963), 'A Comparison of the Social Postures of Some Common Laboratory Rodents,' *Behaviour*, 21, 246-259.

Sales, G. and Pye, D. (1974), *Ultrasonic Communication by Animals*, Chapman and Hall, London.

One of the severe limitations of this kind of study is the absence of marked animals which, of course, renders a considerable amount of information unavailable. Catching birds by net or drugged bait, unless you are fairly experienced, is time-consuming and unproductive – at least in the case of herring gulls, which show a remarkable suspicion of anything even slightly unusual. For the purposes of individual recognition, it may be better to choose a tip which rather few birds visit, and where one or two individuals may be recognized by the presence of a leg ring or the absence of a leg.

Time

This is certainly a project study. A useful project could probably be carried out with 2 or 3 hours' observation a week for 4 weeks, remembering that an hour of taped notes is at least a further hour of transcription; however, the time that could be spent is almost unlimited.

26.3. Farm animals

Purpose

Although the behaviour of farm animals such as cattle and sheep may have been much influenced by domestication, they still show interesting social and other behaviour which can form a useful exercise in field-study technique. Studies in the behaviour of domestic animals are, in fact, lagging behind studies on their metabolism and disease susceptibility, so, apart from being useful teaching exercises, they may lead on to later studies of a more applied nature.

Preparation

In the case of cattle there may be record cards for individual marked animals, giving such things as age, weight, milk yield, genetic relationship to other individuals. It is useful to look at such information before starting observation.

Materials

Per student:
1. A cassette tape recorder
2. Binoculars (about 8 × 30) as required.

Methods

We have no direct experience but it is nonetheless worth while pointing out those areas of investigation most likely to lead to fruitful results.

For cattle and sheep, which have food available constantly, patterns of daily activity may be studied. Records should be kept of the distribution of feeding, ruminating and other activity through the day. Locations of grazing, ruminating and sleeping can be noted. The influence of changed weather or grazing conditions on these behaviours may be observed.

For observing social interactions, cows have the great merit that a number of breeds have patterned hides which make them individually recognizable at a distance. Identification of individuals allows the nature of antagonistic interaction between individuals to be observed and described, and the dominance relations of the group established. The threat and submissive postures in cattle are reviewed by Hafez and Schein (1962). Beilharz and Mylrea (1963a) found a stable but non-linear rank order in a group of 40 dairy heifers. Beilharz and Mylrea (1963b) found that in free movement the group was not 'led' by the most high-ranking animals. There are all sorts of things that could be observed. The relationship between rank order and age, weight or milk yield could be investigated.

In sheep perhaps the most convenient social interaction to study is mother-infant, because lambs remain with their mothers for several months. The duration and frequency of suckling bouts can be scored for lambs of different ages, as well as the lamb-to-mother distance and interactions with other lambs. Ewes do not lose interest in their lambs until they are several months old, so studies on the association between lamb and mother need not necessarily be conducted with very young lambs.

Hunter and Milner (1963) looking at the grazing patterns of sheep on a 250-acre hill pasture, found that individual sheep had a loosely defined home range and that matrilineally related sheep tended to graze the same part of pasture. For someone who is energetic and has access to a few individually recognizable sheep, this would be an interesting line of study.

26.4. Zoo or wildlife park studies

Purpose

In zoos and wildlife parks large animals, particularly mammals, can be observed at close quarters. Many species will be present one or two to a cage, which is of little behavioural interest. It is those species which are kept in groups which can be used in exercises on the social organization of the species. The results of such studies should, of course, be interpreted with caution, since the group may be kept at a density or have an age or sex composition quite unlike the species in the wild.

Preparation

Negotiations should be made with the zoo for observing the animals in comfort and safety. The record cards of individual animals should be inspected to obtain information on age, genetic relationships, etc.

Materials

1. A cassette tape recorder
2. A pair of binoculars (about 8 × 30)
3. A reflex camera with up to a 135-mm lens (for general views of the study area a much cheaper camera would do)
4. A car (ideal as a mobile hide if the study area is a wildlife park situation).

Methods

The study should start with periods of general observation and note taking. For the sake of concentration it is preferable that each period be not more than two hours at a time, but they should be distributed throughout the day to get an impression if certain activities occur more frequently at particular times of day. After this initial phase of observation, a decision should be made on how general or specific the investigation should be and, if quite specific, then what aspects of the behaviour are to be studied. These may well vary, depending upon the animal, but the sorts of results to be obtained may be illustrated by two separate undergraduate studies: one on a group of patas monkeys *Erythrocebus patas* and the other on a group of Barbary sheep *Ammotragus lervia*.

The patas monkeys were a group of 12 animals of mixed age and sex in a wire cage approximately 15 m × 15 m × 6 m. Initially all individuals were captured and marked with coloured dyes; however, after 2 weeks, when the dyes had faded, individuals could be recognized by their natural appearance. A dominance hierarchy was identified in the group in relation to the obtaining of preferred food items. The highest ranking animals by this measure were two adult males. Below these in rank were adult females and other males. The highest ranking animals rarely showed any fighting or chasing, and were responded to by being avoided; however, fights frequently occurred between middle-ranking animals. Infants were seen to associate with their mothers to lesser degrees as they got older, but precise ages were not known. A number of vocalizations, facial expressions and body postures were described, and their functions established by the context in which they occurred and by homology with published work on other primate species.

Jolly (1972) in a useful comparative study of primate behaviour describes patas monkeys as occurring in groups with one or more adult male group leaders and occupying a large home range. The observation in the zoo of frequent fighting was therefore very likely due to overcrowding, and possibly the presence of too many adult males.

The Barbary sheep were a group of 8 individuals in an enclosure approximately 40 m × 50 m. The group was composed of one adult male, four adult females and three juveniles, one male and two female. All individuals were quite easily individually recognizable. The juveniles, whose ages and relationships were unknown, were each observed to associate with particular females, which were therefore presumed to be their mothers. All individuals, except the adult male, were observed to keep quite close together and synchronize very closely their activities of standing, grazing and walking (figure 26.4.1). A particularly interesting finding was that the female who most frequently initiated new group activities was one of the least aggressive animals as measured by the number of head-down threats, rushes and butts directed to other individuals. Females were seen to urinate in response to approaches by the adult male which induced him to *lipcurl* (*flehmen*); this sequence has also been observed by Geist (1971) in mountain sheep, where he interprets it as a mechanism to prevent males approaching non-oestrous females.

There are a number of published papers on groups of zoo animals which are more or less specialized;

Figure 26.4.1 Part of a zoo group of Barbary sheep showing a young lamb resting close to its mother.

however, a paper by Russell (1970) on the red kangaroo gives a good idea of the sort of investigation which could well be undertaken in a student project. Patterns of daily activity are described, together with non-social behaviour such as feeding, drinking and grooming. Among the social interactions described are aggressive and mother-infant interactions, together with their associated movements and vocalizations. The study uses photography to illustrate particular kinds of behaviour and also to obtain line drawings.

Time

This should be treated as a project for one or more students. It will probably require 2-3 hours observation a week for at least a 5-week period.

REFERENCES

Beilharz, R. G. and Mylrea, P. J. (1963 *a*), 'Social Position and Behaviour of Dairy Heifers in Yards,' *Anim. Behav.,* 11, 522-527

Beilharz, R. G. and Mylrea, P. J. (1963, *b*), 'Social Position and Movement Orders of Dairy Heifers,' *Anim. Behav.,* 11, 529-533.

Geist, V. (1971), *Mountain Sheep. A Study in Behaviour and Evolution.* University of Chicago Press (Chicago and London).

Hafez, E. S. E. and Schein, M. W. (1962), 'The Behaviour of Cattle,' in: Ed. Hafez, E. S. E., *The Behaviour of Domestic Animals,* Bailliere, Tindall and Cox.

Hinde, R. A. (1970), *Animal Behaviour: A Synthesis of Ethology and Comparative Psychology,* 2nd ed., McGraw-Hill.

Hunter, R. F. and Milner, C. (1963), 'The Behaviour of Individual, Related and Groups of South Country Cheviot Hill Sheep,' *Anim. Behav.,* 11, 507-513.

Jolly, Alison (1972), *The Evolution of Primate Behaviour,* Collier-Macmillan Ltd. London.

Russell, Eleanor (1970), 'Observations on the Behaviour of the Red Kangaroo (*Megaleia rufa*) in captivity,' *Z. Tierpsychol.,* 27, 385-404.

Stokes, A. W., (1968), *Animal Behaviour in Laboratory and Field,* W. H. Freeman & Co.

Index